P9-AFM-644

Unanimous praise for Charles Bergman's *Wild Echoes* . . .

"A feisty, learned look at America's dead and dying animal species. . . . Beautifully composed natural history . . . simply superb."

Kirkus Reviews

". . . Bergman gives an excellent personal account of his search for encounters with the gray wolf, California condor, Florida panther, right whale, trumpeter swan, and other threatened creatures."

Booklist

". . . Charles Bergman, an environmentalist and teacher whose range is the most immediate . . . Arctic wolves in Alaska, right whales off Sable Island, a solitary dusky seaside sparrow in Disney World in Florida . . . , the black-footed ferret in the Dakotas, panthers and manatees in Florida, and more."

The Boston Globe

"Charles Bergman offers a language with which to read and write and ultimately shape our shifting relationship to the natural world — a language of the human heart, which may help avert the destruction of yet more species in the name of progress."

The Albuquerque Statesman Journal

". . . provocative speculation on the nature of human-wild animal relationships. . . . [Bergman] is a gifted and perceptive writer who consistently compels the reader's attention, even in his most pessimistic moments."

Library Journal

"The author, an environmentalist and professor of English . . . , trekked across North America in search of disappearing wildlife. His work is meditative, mad as hell, poignant, and erudite."

Pacific Northwest magazine

"[Bergman's] history of man's perceptions of animals and nature is impressive and valuable to our understanding of the issues . . ."

Seattle Weekly

WILD ECHOES

Encounters with the Most Endangered Animals in North America

Charles Bergman

Alaska Northwest Books™

Anchorage • Seattle

For Ian,
   the wizard of fire

and Eric,
   the genius of smoke

Copyright © 1990 by Charles Bergman
All rights reserved. No part of this book may be reproduced or transmitted in any form or by any means, electronic or mechanical, including photocopying, recording, or by any information storage and retrieval system, without written permission of Alaska Northwest Books™.

Wild Echoes: Encounters with the Most Endangered Animals in North America was first published in the United States by McGraw-Hill Publishing Company, New York, 1989.

Library of Congress Cataloging-in-Publication Data
Bergman, Charles.
    Wild echoes : encounters with the most endangered animals in North
  America / by Charles Bergman.
      p. cm.
    Originally published: New York : McGraw-Hill, © 1990.
    Includes bibliographical references and index.
    ISBN 0-88240-404-0
    1. Endangered species—North America   I. Title.
QL84.B47   1991   90-45010
591.52'9—dc20   CIP

Cover: A gray wolf (*Canis lupus*), photographed in Denali National Park. (Photo by Ron Sanford, AllStock)

Cover design by Alice Merrill Brown

Alaska Northwest Books™
A division of GTE Discovery Publications, Inc.
22026 20th Avenue S.E.
Bothell, Washington 98021

Printed on Acid Free Paper

Printed in U.S.A.

CONTENTS

ACKNOWLEDGMENTS

I am indebted to many people for their help in making this book possible. In my discussions with them, over matters directly related to the text of this book, or over more general ideas that are the context of this book, I have learned how important conversation is to discovery.

Many people helped me in the field, supporting my efforts to find the rare animals described in this book and often providing company, meals, or a place to stay. Anyone who has traveled in ways that make him or her dependent on others knows the kind of appreciation I mean to express here. I can honestly say I have enjoyed all the people I met, even when I might have disagreed with them on philosophical issues.

In the text, I have described many of these individuals. There are many other people, not mentioned, whom I would like to acknowledge.

Art Wolfe and I often traveled together. I admire his superb eye for nature and his outrageous, funny wit. And to Scott Maloy, with whom I also camped and traveled through some remarkable back country, thanks for all we've shared.

A number of people contributed to my work with wolves: Rolf Peterson on Isle Royale, Joel Bennett and Vic Van Ballenberghe in Juneau, Warren Ballard and Wayne Hall near Anchorage, and Lu Carbyn in Alberta.

To Wes Biggs, who threatened to make me "dogmeat" if I didn't mention our unbelievably good time birding in south Florida: thanks for your knowledge of animals, as well as the limpkins, king rails, snail kites, short-tailed hawk, and much more. Sorry you got sick.

James Wiley helped me in Puerto Rico after I thought everything had gone wrong, and Angela Arendt's meal of paella and tostones did me a world of good after the horrors on the coastal plain. Thanks also to Jan Kalina and Tom Butynski—I'm sorry I haven't seen you with your gorillas in Uganda yet. Maybe someday.

George Horsecapture moved me with his talk of quests and visions in the Buffalo Bill Museum in Cody. Thanks, too, to Jeri Reeves for her kindness. Emily Hill, also, in the nursing home, left me speechless when she said, "You came too late. All gone."

For his devotion to trumpeter swans, as well as all waterfowl, I have enormous respect for James King in Juneau. Thanks also to Terry McEneany and Martha Jordan. And to Jack and Trudy Turner, at Lonesome Lake, British Columbia—where my life was changed more than they could know.

For a wonderful time on the misnamed Lucky Seven, I want to remember Martie Crone, Amy Knowlton, Greg Stone, and Laurel Code. And Pip, too, who loved fixing quiche and salt cod soup. Brian Hoover, who died on Mt. Everest before he could go searching for monk seals in the Caribbean—we all remember him getting new red Converse tennis shoes after the plane went into the Atlantic off the Florida coast.

Tom Stehn showed me whooping cranes in Aransas and on Matagorda Island, along with much more. Eric Forsman has always been very generous with his time and knowledge, indulging me in one of my favorite obsessions—spotted owls.

Both John Ogden and Tim Clark also shared their knowledge about condors, ferrets, and conservation issues.

Many people devoted their expertise to portions of my manuscript, helping to spot inaccuracies and offering responses to the ideas. Their help was invaluable: Robert Stephenson, Herb Kale II, Wes Biggs, Noel and Helen Snyder, Judith Delaney, Pat Hagan, Dave Maehr, Melody Roelke, Tom Thorne, Scott Kraus, and Jerome Jackson. They have made the book better. I am responsible for any errors that might remain and for the ideas that make up their context.

To the magazine editors with whom I have worked on wildlife stories, I want to express my gratitude for their intelligent responses to my texts and their support for my research: Mary Smith and Andrew Brown at *National Geographic*; Les Line, Gary Soucie, Martha Hill, and Roxanne Sayre at *Audubon*; Marlane Liddell and Don Moser at *Smithsonian*; Eric Harris and Ross Smith at *Canadian Geographic*; David Brewster and Rose Pike at the *Seattle Weekly*.

To Michael Marcotte at KPLU-FM, in Tacoma, Washington, I want to express my thanks for the opportunity to do commentaries over the radio. That has been a wonderful experience, and I think it helped my writing in many ways.

Tom Miller, my editor at McGraw-Hill, gave me advice and encouragement through all stages of the writing of this book, from that first breakfast long ago. I want to thank also Lisa Frost, my editor through the final stages of editing and production at McGraw-Hill.

My agent, Sherry Robb, not only helped make this book a reality but also, through her enthusiasm and confidence, often gave me energy.

My thanks also to Pacific Lutheran University for its support in faculty research grants. Lennie Sutton was a magician in the computer center, infinitely resourceful in working with the manuscript. The librarians were always cheerful in helping me track down sources.

Several people have read the manuscript in various stages of completion, and their ideas, encouragement, and criticism were more important to me than I can say. Dan Morgan and Paul Webster read early drafts, and have been steady friends through many years. Kay Garcia gave me wonderful encouragement. Marilyn Davie also read an early draft, but her help in this project goes far beyond any single response to a text.

I want particularly to thank Lynne McGuire and Sandy Faye-

Cassio for their reading of my later drafts: Lynne—smart, tough, with an edge of cynicism, responsive to ideas, and candid; Sandy—smart, enthusiastic, learning to write herself, struggling too with questions of voice and stance. Carol Allen offered very good suggestions, and Marie Wutzke gave a close reading when I needed one.

Several people helped in less specific ways, and I must thank them. Robin Brown, at the University of Minnesota, has long stimulated me with theoretical perspectives on language, reading, and nature, and they continue to open my eyes. Elaine Maimon, at Queens College in Flushing, has more than any other person helped me believe that I could write. I owe to her a whole new view of the writing process, one that has encouraged me to believe that I could find what was in me to say if I had the courage to write what I think.

To a group of friends, I am grateful for their support during the writing of this book: Irv Goldberg (whom I miss), Gerry McCarty, Mike Dash, Barb Bennett, Terry Lull, Terry Hammerly, Mike Winter, Gordon Church, Nancy Witt, Karen McLain, Helen Saum, Marlene McFarlane, Dave Vicks, and Sandi Belloti.

Three men at Pacific Lutheran University have given strength of varying kinds: David Seal, who shares my interest in animals and whose ideas and language are always stimulating; Richard Jenseth, a rhetorician who has a remarkable courage to be who he is; and Paul Menzel, alias Apollo, who has been a brother to me. They have all contributed, without necessarily knowing it, to this book.

For that which befalleth the sons of men befalleth beasts.
Even one thing befalleth them: as the one dieth, so dieth
the other; yea, they have all one breath, so that a man hath
no pre-eminence above a beast; for all *is* vanity.

Ecclesiastes iii, 19

It is neither spring nor summer: it is Always,
With towhees, finches, chickadees, California quail, wood doves,
With wrens, sparrows, juncos, cedar waxwings, flickers,
With Baltimore orioles, Michigan bobolinks,
And those birds forever dead,
The passenger pigeon, the great auk, the Carolina paraquet,
All birds remembered, O never forgotten!
All in my yard, of a perpetual Sunday,
All morning! All morning!

Theodore Roethke, from "All Morning"

INTRODUCTION

I have a fantasy: I imagine myself sitting in my living room, on my sofa. Outside my house—outside *our* house—animals are gathering. Lost animals, endangered animals. Peering in the windows, with strange expressions on their gaunt faces. Murmuring indistinctly of Something Else. Murmuring in wild echoes.

It's a strange feeling, knowing the animals are near, even as we think we've distanced ourselves from them. They give me the sense that there's more to life than I have known before—the sense that these animals know things that I don't know.

This book is my attempt to understand, in both social and personal terms, the meanings of the phenomenon of endangered species. I have come to view endangered species as another of the great topics that challenge our certainties, like dreams and madness, sex and death, and the poor. These broken creatures, haunting the margins of our lives, are less a part of nature than of our culture. We made them, we preserve them. Ironically, they are one of our totems; they are the true animal emblems of our culture. Extinction, like the poor, may always have been with us, but endangered species are a modern invention, a uniquely modern contribution to science and culture. They are one of the unhappy consequences of the way we have come to know animals, the dark side of our relationship with nature.

Yet, in their imperiled status, endangered species represent a paradox: Though they are the result of our long obsession with power over nature, they embody the limits of that power. They are a mirror, not of our stunning triumphs over nature but of our failures. Regardless of what we say about our feelings for animals, endangered species and the epidemic of extinction that are so closely associated with Western culture speak a strict and severe language. They are images, reminding us that the more complete our domination of nature has become, the more we have lost in the process.

The creatures in this book have changed the way I see wild animals, and they have helped change the way I see myself. One experience many years ago suggests how endangered animals have changed my feelings—not just for nature but for life. These feelings, evoked by a spotted owl in the state of Washington, have been the impetus behind this book.

With several friends, I was backpacking through an old-growth forest along the boundaries of North Cascades National Park.

The ancient forests of the Pacific Northwest—what little is left of them—are an experience in natural gothic: trees surging to invisible tops, weird shapes of knotted trunks and twisted branches, logs rotting on the ground, views dissolving in shadowy distances. These forests, always darkly green, are sometimes more than 400 years old: cedars whose branches slouch in layers of casual droop; lacy hemlocks with supple lines and a graceful curve to their topmost branch, swaying in the slightest wind; stiff-needled Douglas firs.

It was a day of intermittent sun. Yellow light sliced through the green shadows in slanting shafts, the dust careening through the beams like atoms. We hiked through an immense stillness, a heavy, wet, and muffled silence.

We talked little, our private reveries interrupted only occasionally by the ebullient melody of a winter wren or the lisp of a chestnut-backed chickadee. Hot and tired, we reached a fork in the trail after several hours of steep switchbacks, dumped our packs, and collapsed in the cool shadows, lounging in a vaporous forest.

I was bolted out of a half-dream, as I lay there, by the unmistakable call of a spotted owl.

Jumping to my feet, exhaustion and lethargy forgotten, I searched all the gargoyle shapes in the trees around us for the owl. Nothing. Slumped and comfortable, my friends laughed at my sudden energy, my absurd burst of inspiration from a single note. In spite of them, I hooted back into the forest, imitating the owl as I rushed up and down the trail trying to find it, craning my neck in futile contortions to stare into the recesses of the canopy. After a while, I gave up, and we started hiking again.

But there, only about 50 feet farther up a steep bend in the trail, I found the owl sitting on the low limb of a Douglas fir, cool and impassive as a Tao philosopher.

We stood on the trail, the four of us, gawking at this owl for well over half an hour. With its chocolate-rich, white-dappled plumage, it melted into the dark shapes and checkered shades of the forest, watching us with an expressionless aloofness, self-contained, revealing nothing, tolerant but obviously unimpressed. When we moved, its head swiveled sharply and it stared right into us. A spotted owl's eyes are huge, and the deep, concentric rings that spin around them exaggerate the effect. They suggest the intelligence and inscrutable wisdom that are part of what owls have always represented for us. Most owls have yellow eyes, with a possessed intensity in their stare; trying to meet their gaze is like looking into brass cannons. But a spotted owl's eyes are black and glimmering, with a softness to them that suggests a kind of approachability.

I have always loved owls. They are as close as anything I have to a totem. None of the nineteen species of owls in North America is on the official federal list of threatened or endangered species, but the spotted owl should be. Its Latin name is *Strix occidentalis.* It has been nominated for listing, and at the rate that the ancient forests in the Pacific Northwest are being cut, the spotted owl's elevation to the status of one of our country's premier endangered species is merely a matter of time and inevitability. Estimates vary—they always do when we try to determine the aggregate impact of our actions on the environment—but there is as little as 5 to 10 percent of the original old-growth forest left in the Pacific Northwest, with plans for cutting more. The spotted owl depends upon these huge trees—it feeds on the flying squirrels that live in the old growth, and it nests in the cavities that form in the dying trees. As if waking slowly from a deep sleep, we are beginning to realize how precarious the status of the spotted owl is and just what some of the implications of our clear-cutting timber practices are.

Staring at the owl, and explaining some of the significance of its status to my friends, I realized that my happiness at finding it contained also a sharp edge. I felt an intense, painful desire for *more*, though I could not tell you what I wanted more of. The yearning was deep and explosive, a force straining to break loose, an urge to grab and hold onto this life sitting above us—precisely because I could feel it slipping away even as we looked at it.

It is this sense of the vulnerable and fleeting that has become part of my experience of nature—of *our* experience in nature. And the sense of imminent loss has, ironically, become a basic part of the value of nature in our times. Too often, the fact that an animal is endangered or threatened constitutes the explanation of why that creature is important. We like animals *because* they are endangered, as if being endangered confers prestige and status. There is something troubling in this for me. Except in a kind of naïveté, it is no longer possible to return to nature, as Shakespeare's characters did or as Thoreau did, expecting some kind of renewal of purity. The innocence of nature is gone, and the epidemic of endangered species is one measure of just how far we have come from that innocence.

What I discover, as the old dreams of nature fade or are shattered, is a new and more complicated experience in the wild. Even as I enjoy the beauty of the moment, the sense of involvement in the world, I feel the presence of imposed danger, the sense that this animal, this owl, is just a thin echo of what once was. Or worse, that it may soon fade even further and vanish. Because if endangered animals are disembodied voices of a former abundance, extinction is a soundless echo.

I try to imagine Adam, in the classic Western image of a human responding to animals, sitting beneath the tree of knowledge, watching the bizarre and glittering array of creatures that parades before him. He gives each of them a name, and there is no creature that he calls endangered. The scene is an intimate one. For Adam, this was a moment of introduction, a testimony to the abundance of God's creation, a promise implicit in the act of naming that the creatures were knowable. It was also a sign, in the naming, of Adam's superiority to the beasts. Now that very superiority—found not only in this biblical tableau but also in the mechanistic assumptions at the core of our science of animals—has turned on us. It promised dominion over the beasts of the field and power over nature, yet it has left us isolated and alone. We live, as Ezra Pound said, "in an age of science and abundance." But we are surrounded by images of loss.

The modern pageant of endangered creatures contrasts sadly with the image of Adam under the tree: Moments of introduction and discovery have been replaced with farewells. Animals vanish even as we learn their names for the first time, creatures most of us will see only in photographs. Like the spotted owl, our animals parade before us, not to enter our lives but to leave them.

We are living at the ends of traditions. In endangered animals, I hear some of the rumblings of my own discontent.

The environmental movement, and the effort to save endangered species, is founded largely on the theoretical essays of Aldo Leopold. Though the movement is older than Leopold, he gave it both ground and impetus in 1948 with his justly famous essay "The Land Ethic." In it, he argued that we need to see the land—which for him was the interlocking web of animals and plants and soil—as a community of which humans are a part. "In short," he wrote, "a land ethic changes the role of *Homo sapiens* from conqueror of the land-community to plain member and citizen of it. It implies respect for his fellow-members and also respect for the community as such." We now popularly call this concept "ecology." Aldo Leopold was part of the great effort to break down the ancient Western dualism between humans and nature. He showed us that we have a place among the beasts.

But this concept is limited by its literalism. The land ethic, and the environmental movement built upon it, still identifies nature as something *out there*, as something external to us, something separate. We need a new, more intimate concept of the ecology of people and animals, humans and nature. If we look upon animals as symbols, as entities partly created by and mirroring us, we can see another dimension of our relationship with them: the emotional and unconscious meanings we attribute to animals. This approach shows how nature is created by culture. What we have not yet learned to understand, still living as we do in the outer light of an empirical and mechanistic age, is the place of the beast inside us.

What forms of consciousness lie behind, and determine, the structure of our relations with animals? How is it that we know these animals, and how are we always involved with the animals by the very act of knowing them?

The endangered creatures in this book are not merely problems to solve, as if we could finally figure out just what to do about their predicaments. Rather, they are perpetual questions. They are voices from the Other, external images of what is alien and foreign inside us, silent and shadowy strangers. I do not for a minute presume to think I can ever know these creatures completely. But I do think more complete relationships are possible, full of richness and emotion. Endangered species are not simply accidents of our way of living. They are the necessary consequences of our way of knowing animals. Endangered species reveal some of the rifts and blank spaces in our ways of seeing, and in those rifts, if we are willing to pay attention to them, I see the possibilities of new forms of knowing, new ways of feeling.

On the hike into the North Cascades, one of the friends who stood beside me below the owl was a philosopher, a specialist in ethics at the university where I teach. As I talked about the owl's rarity and the controversy over protecting it from timber interests, he posed the fundamental question.

"Why," he asked, philosopher that he is, "why should we preserve this owl? In fact, why preserve species?"

The question made me mad. It still irritates me, no doubt because there is no satisfactory answer.

But I now think it's the wrong question.

Built into the question "Why save endangered species?" is all the arrogance of centuries of Western domination over nature. The question presupposes that we are the lords of creation, that it is our right and our duty to oversee nature. It also exposes our limited view of nature, even in our concern for it: Animals are something for us to control.

The best answers speak of the intrinsic value of life—the life that we share with all of this creation. They also speak of compassion, which is the philanthropist's virtue. Yet compassion can itself be a kind of arrogant emotion, a condescension: From our exalted position, we will take pity on the less fortunate creatures, the ones unable to adapt to life on the terms we have set—fast-paced change and sweeping destruction of the earth, both of which nullify hopes for evolutionary adaptation. Compassion is noble, and certainly better than dismissing the creatures, but compassion does to the endangered animals what it has done to the poor: It turns them into moral failures. They aren't as good as we are, we think smugly, not as strong or powerful, and thank God we're not like them—but we'll give them a hand anyway.

The worst reasons for saving endangered animals are probably the most effective: the ones that are blatantly, distressingly utilitarian. Although they galvanize the most support for saving species, such reasons are just one more excuse for exploiting animals, for predicating their lives on our needs. We should save endangered species, so the argument goes, because we never know where the cure for cancer will come from.

In the politics of preservation and extinction, these arguments have not really worked. The United States has undertaken the most progressive and ambitious attempt to save endangered species that any country has ever made. In 1966, we passed the Endangered Species Preservation Act, and in 1969, the Endangered Species Conservation Act. Both were limited in comparison to the Endangered Species Act of 1973 (amended twice, both times weakened), which requires that all federal departments and agencies protect endangered species and their habitats. It prohibits the taking of and trade in endangered species, with some exceptions. In addition, we passed the National Marine Mammal Protection Act in 1972, which placed a moratorium on the taking or importing of marine mammals like whales, seals, otters, and manatees.

This protection has resulted in notable successes. The whooping crane, a majestic white bird with one of the most mystical of mating rituals—a gangly and eerie wild dance—is one creature saved by our attention. Its Latin name is *Grus americana*. The tallest North American bird at 4½ feet, with a 7½-foot wingspan, the whooping crane was once down to as few as twenty-one birds. As of 1987, there were 134 whoopers returning to the wintering grounds of Aransas National Wildlife Refuge in Texas and making the long and hazardous flight back to Wood Buffalo National Park in northern Canada, where they nest. Both nesting and wintering areas have been set aside expressly for the cranes.

Yet even this success is largely symbolic. Once there were perhaps 1400 whooping cranes in North America, and before we rush to self-congratulation for saving the whooping crane from extinction, we should recognize that the place we have saved for the cranes is marginal at best, and always in jeopardy.

More worrisome: While we have saved remnants of some of the more spectacular species of the last few decades—peregrine falcons, bald eagles, and brown pelicans, among others—the problems facing our wildlife continue to escalate.

As of May 1988, there were 495 U.S. species or subspecies either threatened or endangered on the federal list (377 endangered; 118 threatened). Making the list is the crucial achievement for any species in trouble. But the backlog of species nominated for listing, yet getting lost in the political and bureaucratic entanglements of approval, has remained constant for over a decade: between 3000 and 4000 species. As in the case of the spotted owl, nothing can be done on behalf of these species, except public agitation, until they are listed. They languish in the shadows of our attention.

But listing is no guarantee of recovery. Of the 495 species, only about sixteen are recovering. Another eighteen species are now thought to be extinct, according to a Government Accounting Office report. (The eighteen probably have not gone extinct since the passage of the Endangered Species Act; rather, scientific opinion now considers that some may already have been extinct before the passage of the act.)

Our approach to the problem of endangered species has been a classic example of American pragmatism: Define a problem in terms that enable us to *do* something, and then invent a solution. But the problem of endangered species has not been easy to contain or manage. While we have focused on particular species—high-profile animals, primarily—the problem has continued to grow and has reached global proportions.

Determining extinction rates is a tricky business, really amounting to no more than astute guesses. But since the rise of modern culture in the seventeenth century, the rate of extinction worldwide has climbed to unprecedented levels. According to Paul and Anne Ehrlich, "The rate of extinction of bird and mammal species between 1600 and 1975 has been estimated to be between five and fifty times higher than it was through most of the eons of our evolutionary past. Furthermore, in the last decades of the twentieth century, that rate is projected to rise to some forty to four hundred times 'normal.' "

At best, we have achieved important but limited successes, small refuges of survivors amidst a storm that continues to howl all around. Somehow, the question "Why save endangered species?" presumes that animals are separate from us and that it might be all right to lose them. It presupposes that endangered species are a given, something we have to live with, one of the "data" of our lives, to use the technical language that is so much a part of our times. It is a rationalist's response to animals, and it derives from the same sensibility that has created the problem of endangered species in the first place. In this context, work on behalf of the animals is always a rearguard action, and we are reduced to well-meant but ultimately feeble gestures, such as putting tigers in zoos to protect them. Meanwhile, the forces that have created the problems surge on unabated; even legislation on behalf of animals is contested or diluted or manipulated, and the wreckage worsens.

The question I want to ask is: What do endangered species mean? Related to this question are others: Why is it that *we*, in our culture, have this new epidemic of lost animals? What does it say about us and the meaning of nature in our culture?

My intention in this book is to raise questions that will lead to a rethinking of our relationships with animals, questions that will open the possibilities of basic change. One of the premises of this book is that animals are only partly biological creatures. They are also symbols in which we can read who we are. When we look at endangered species, we can learn not just about animals but about ourselves.

So I want to turn the equation around, and see not just what we have to say about the animals but what these animals have to say about us. By asking how we see animals, and why we see them as we do, we can gain a new perspective on our relationship with animals. Perhaps also we can re-create animals, by learning to re-imagine them and our world. In each of the following chapters, I focus on a particular endangered animal as a representative, a crucible of loss and possible recovery. Taken together, the chapters are about our need for a new spirit in an age that is consuming itself, dying on its own ashes. They are, as it were, my attempt to understand the most mythic of endangered species—the phoenix. Surviving alone, the phoenix was the fabled bird of the soul, thought to die in flames at the end of each millennium, only to be reborn out of its own ashes, heralding a new age.

ONE

HUNGER MAKES THE WOLF

I

The wolves had dug their den into the crest of a small ridge in a dense stand of spruce trees. We meandered toward the den site, keeping to the thin line of one of the many wolf trails, past several lifeless beaver ponds and through the wet underbrush. High in the Alaska Range, the soggy clouds hung off the sides of the surrounding peaks like wet clothes, loose and heavy. We emerged from the dark spruce into a clearing where the wolves had trampled the dirt around their den into a hard pack. A small hole in the ground, the opening of the den looked like a dirty mouth puckered into a belch, and around the mouth was scattered the gnawed litter of former meals —four beaver skulls from the nearby ponds, moose bones well chewed by teething pups, an antler, a ram's horn from a Dall sheep, duck feathers.

Though the wolves were gone, and though there was a calm in the darkness of the shaded den site, it took little imagination to feel the intimations of ferocity and passion in the marks of the fangs on these bones. I found myself thinking not of comfortable domesticity but of Francisco Goya and his painting *Chronos Devouring His Children*—an image of savagery and hunger. And I realized for the first time, that one aspect of living in time and nature is to live with hunger.

Carl Jung associated Saturn, the Roman name for the Greek time deity Chronos, with the *Sol niger* of alchemists, the dark or shadow side of the sun, and claimed that the ancients knew this destructive aspect of the sun. That's why Apollo, god of the sun, was also associated with mice and rats. And with wolves.

I found an image of wolves at this den site that could not be easily domesticated, some threat always hovering on the edges of our lives, on the borders of our consciousness. The wolf has long been an image of this frightening hunger, both in nature and in the human spirit. In *Troilus and Cressida*, Shakespeare looked inside and found "appetite, an universal wolf," that devours "an universal prey."

Instead of facing this dark aspect of the wolf and other animals, we have in this country and this century worked to domesticate nature. Both scientists and environmentalists have contributed. Scientists, by working to reduce animals to rational and objective terms, to numbers and formulae and laws; environmentalists, by making nature into a mirror of our middle-class, comfortable culture, a realm of tidy homes and nurturing families.

In both cases, we remain alienated from some deep and primal power in the wolf, a power not accessible to reason and not necessarily conformable to culture. As I stood by the mouth of the wolves' den, I felt stirrings of this power.

I had come to this wolf den with Danny Grangaard, a biological technician for the Alaska Department of Fish and Game. We had been helicoptered deep into the Alaska mountains, to a place too remote and rugged for a regular plane. Danny was taking part in a long-term study to gain a scientific understanding of the relationships between wolves and their prey. Shortly after we walked to the den, Danny slipped on rubber gloves and began collecting the bones and other litter in a plastic garbage bag.

I let myself go with my feelings.

The wolf has always evoked powerful passions in humans. Very few Americans, for example, have ever seen a wolf, yet the animal lives in all our imaginations. Every animal wears a double face: It is both creature and symbol, fact and analogy. Seeing a wolf as an animal, as Danny was trying to do in this study, tells what the wolf *is*: It helps us see the wolf rationally, objectively, with a certain kind of distance. Seeing a wolf as a symbol tells us what it *means*: It helps us see the wolf imaginatively, often with a confusing intimacy.

The symbol shows the wolf that lives inside, and much that it reveals is not only cultural but personal. It connects us to our childhood dreams, to our most basic needs and desires, and to the demonic.

In a strange way, the images of hunger around the den did not frighten so much as exhilarate me. Looking at the mouth of the den, I could feel the wolf inside. Like Alice in Wonderland, I had an impulse to crawl in and see the pups—an irresistible urge to see the wolf face-to-face, to confront the beast in its den.

Danny said he thought it would be all right. I grabbed a flashlight and dropped to my hands and knees. The hole was only 14 inches wide. I could barely fit through. Still, I wormed my way in up to my hips. The burrow dropped straight down and then angled hard right. It was crowded inside; my arms were cramped, I was barely able to reach ahead and shine the light. The dirt walls were wet, dripping, stinking of wolf piss and damp fur. In the beam of the flashlight, about 7 feet away from me, three wolf pups cowered in the far corner of their den, a protoplasmic pile of bluish-gray fur.

One brownish ear, I remember, was crushed against the top of the den. A couple of huge rust-colored paws poked out from the pile, precise owners unknown. The pups had small snouts on adorable faces, giving only faint hints of the fangs to come. Their eyes reflected a dull blue from the flashlight, expressionless and completely impersonal, inhuman.

The pups growled at me in low, menacing tones, the snarling edge of the unknown. Those throaty growls were the fiercest threats I've ever heard.

This was as close to the pure experience of the alien and the wild as I have ever come in nature, as raw as the bones the wolves themselves had chewed. These wolves were untamable. Facing them in their den, I found myself at the borders of the irrational, staring in awe at its power.

The wolves in the den—a world beyond reason and culture.

I thought later of one of La Fontaine's fables, of the wolf that preferred a life of hunger in the wild to the well-fed life of a dog wearing a collar. At some barely accessible level, this is the wolf that lives in the American psyche, part of the fabric of our daily language:

Hunger makes the wolf.
We wolf down our food.
The wolf at the door.
We cry wolf.
A man who loves women and sex? A wolf, of course.

These images unconsciously shape the way we see wolves; most of them are full of fear, defined by hunger and sex and a sense of intrusion. Fantasy is always a precondition for our perception of reality. The wolf in our imagination describes an interior landscape of fearful and alien impulses, projected onto the beast beyond.

But there is more to this wild wolf.

When we are crazy, on the edge of madness or uncontrollable passion, then what? In that irresistible urge for freedom and that unrestrainable wildness, we howl with the wolves at the moon.

Such wildness wears a horrifying mask too. One manifestation is lycanthropy, which once was considered a medical condition. Overcome by melancholy, a person could be transfigured into a wolf; stealing into cemeteries and churchyards at the dark hour of midnight, the afflicted person digs up corpses like a wolf. It is a gruesome association of death and hunger and the travesty of the sacred.

The mistake is to take all this literally. It is psychology. A man-wolf, or werewolf, might howl fearfully, but the fur is not really on the outside. The more horrifying fur grows beneath the skin, creepy and itching, felt invisibly.

Here—in the human soul—is the second true wilderness. Animals like wolves can give us some of that wild energy that William Blake called "the opposite of reason." The wolf den, and the experience of crawling into it, evoked for me this deeper layer of meaning. We all have the right to construct our own meanings out of our experiences in nature: Nature is not only an objective world but also a world we create.

More is at stake in saving a wild wolf than we realize. We need to save both the literal and the symbolic creatures. Part of the reason so many animals are endangered, in this rational and empirical age, is precisely that we have forgotten the two faces of every creature. We have forgotten what animals *mean*.

I don't want to settle for only half the animal. Already, we are learning to accept diminished beasts and diminished selves. In facing those wolf pups in the den, I entered nature in a new way, engaged and passionate. My experiences with endangered animals have become for me an ongoing struggle to discover new ways of perceiving these creatures, to respond more immediately to animals, and to explore a much wider range of possibilities in living.

After we finished at the den site—after I crawled back out of the den and Danny gathered up all the prey remains—we hid in a blind in the woods. We hoped to have one of the rarest experiences in wildlife study—seeing the parent wolves with their pups at the den. I wanted especially to see the mother.

II

Endangered animals are, largely though not completely, the victims of our culture, martyrs of the cult of progress. Each species has traced its own narrative toward its precarious status, and the particularities of its story are determined mostly by the nature of its relationship with our culture. Like individual people, each species has its own vulnerabilities, its own ways of suffering. And you feel, after studying endangered species, that you are growing expert in the permutations of inflicted suffering, a kind of displaced, biological sadism.

If the animals themselves weren't so wonderful, so compelling to seek and so exhilarating to encounter, following their stories would be like documenting a disaster. The scope of the enterprise would be narrow, like medicine reduced to the specialty of autopsy.

The wolf in North America is one knot in the weave of our relationships with wild animals, and it shows us one way that science has taught us to see and treat animals.

When I came to Alaska, I had joined a number of biologists working on wolf research for the Alaska Department of Fish and Game. Danny Grangaard, who was with me at the wolf den, was on one of the research projects around Tok, Alaska, near the Canada-Alaska border and the Alaska Range of mountains. The point of the research, and the reason that Danny collected the bones by the den, is to understand the effects of wolves on populations of moose, caribou, sheep, and other prey. In more technical terms, the research is on predator-prey relations.

The research on wolves is both very sophisticated and extremely controversial. The controversy stems from the fact that the research does not simply satisfy the theoretical curiosity of the researchers. For the last decade and a half, the wolf research in Alaska, Canada, and Minnesota has also been used to justify "wolf control," the official euphemism for government-sponsored killing of wolves to manage their populations.

The larger question this research raises is how pure or objective wildlife biology can be in the understanding it gives us of animals, in this case of wolves and their prey. Put another way, what is the intersection here of science and politics, objective knowledge and self-interest?

Everything about wolf research stirs fierce emotions. Beneath the passions and the politics, the wolf as an animal remains elusive, with matters as basic as its North American population debated intensely among experts. Yet the wolf has been studied intensively in North America for almost half a century.

Wary and intelligent, *Canis lupus* is called both timber wolf and gray wolf. The pelt color in the species varies widely, from almost snow white to black. Usually, an individual wolf's coat blends several shades of fur—grays, rusts, blacks, whites. The wolf is found throughout the northern hemisphere and is considered by most experts to be a single species, though there are twenty-four recognized subspecies in North America and eight in Europe and Asia. The rare red wolf, which is currently being reintroduced to the wild in the southeastern United States, may be a variant of the gray wolf, though opinions on that differ.

In the contiguous United States, the gray wolf survives only in vestigial populations. The largest remnant, about 1000 to 1200 animals, lives in the wilderness of northeastern Minnesota, in and around the Boundary Waters Canoe Area and Voyageur National Park. Some of these wolves have wandered over into Wisconsin, where there are now about twenty or thirty wolves. Mainland Michigan supports a half dozen or so wolves, mostly lone wolves; transplant efforts into the state have been largely unsuccessful. Currently, about twelve wolves live and roam on Michigan's Isle Royale in Lake Superior, a number that fluctuates up and down over the years. About a dozen haunt the forests of Montana, and proposals to reintroduce the wolf to Yellowstone National Park continue to arouse anger on both sides of the issue. All the wolves in the contiguous United States are protected by the Endangered Species Act.

In Alaska and Canada, even the most basic aspect of wolf biology is an area of contention. The current official population estimate of wolves in Alaska, for example, is 6000 to 10,000. Such figures not only mean the wolf is not in immediate danger of extinction but also are used to show that the population is healthy and that wolves can be shot, or "controlled." Critics of these numbers, and the census that produces them, claim that the wide range in the estimated numbers is itself a sign of their dubious value, rendering them unreliable at best and, some say, meaningless. Canada gives an official estimate of 50,000 wolves, but since the country has no formal census for wolves, the figure is an extrapolation, and some call it a guess.

Despite a range of attitudes, the majority of wolf biologists find the evidence compelling: Wolves, they say, can and do decimate the populations of their prey, especially deer, moose, and caribou. A number of factors seem to be involved in several recent drastic declines—the biologists call them "crashes"—of prey species, especially severe winters, overhunting, and predation. The debate among biologists hinges on how to weigh the impact of each of these factors on, say, a region's moose population. Those conducting the studies feel they have isolated the wolf as a main culprit in many regions, and they maintain that wolf control is guided by their research.

In 1975, arguing that it would help to restore depressed populations of moose, caribou, and deer, game managers in Alaska resumed a discontinued program of killing wolves. In the winter of 1977–1978, for example, 920 wolves were killed by trapping and aerial shooting. (This figure represents the total "harvest." Fewer than 150 wolves were taken in the "control" program.) In Canada, three of the nine provinces conduct wolf-control programs. In British Columbia, government agents in 1983–1984 shot 330 wolves; using planes and helicopters to locate the packs, the agents killed the wolves from the air.

The killings have provoked mountains of hostile letters, public protests, and court actions that still continue.

Until three decades ago, the federal government had paid hunters to kill wolves, but the wolf bounties were dropped in Alaska in 1959 and the numbers of both wolves and prey increased dramatically. Biologists for the Alaska Department of Fish and Game thought the situation was reaching a stable balance. Then a series of severe winters hit Alaska in the late sixties and early seventies, and wolves crashed in numbers along with their prey.

In order to understand what was happening, Bob Stephenson and several colleagues in the Alaska Department of Fish and Game began what was some of the most important research on wolves and their prey. I met Stephenson in Fairbanks, a cheerful, friendly man with a great sense of humor, and joined him several times in the field. According to Stephenson, their studies forced them to rethink the relationship between wolves and prey. Their main study area was in the Tanana Flats. During the devastating winters of the late sixties and early seventies, moose had dropped from a previous high of about 20,000 in 1965 to a low of 3000 in 1975. Caribou numbers crashed, too, from 5000 in 1970 to 2000 in 1975.

Most of us have cut our ecological teeth on the concept of the balance of nature, a concept enshrined until recently in graduate schools of wildlife management. The notion is that predators and prey live in a delicate harmony, exquisitely sensitive to each other.

They form a kind of contained system, with self-regulating mechanisms to keep their populations healthy. On the basis of this theory, researchers like Stephenson expected the populations of moose and caribou, as well as of wolves, to rebound vigorously.

They did not, despite a succession of mild winters after the harsh winters. The biologists were forced to rethink what was going on.

Stephenson acknowledges that one pressure was overhunting. But he argues that in 1973 the state took action by closing the caribou season to hunters in the area, and it squeezed the moose harvest down to less than 3 percent of the herd. Nevertheless, according to Stephenson, biologists were surprised that the populations of ungulates (hooved mammals such as deer, caribou, and moose) did not rebound. In this particular area of Alaska, a wolf pack will take a moose every five days or so. To the biologists in Alaska, the wolf seemed to be one reason for the low numbers of moose and caribou.

They decided to test that hypothesis. Each year between 1975 and 1979, the Alaska Department of Fish and Game experimentally killed 38 to 60 percent of the wolves in the Tanana Flats by trapping and shooting from planes. Compared with a control population in the Nelchina Basin near Mount Denali, moose and caribou calf survival jumped twofold to fourfold almost immediately. Today, moose populations in the area are high. This experiment and similar experiences elsewhere in North America convinced the people in the Alaska Department of Fish and Game, and many of the other biologists I have met, that wolves can limit the numbers of prey.

To use biological language, the wolves create a "predator pit," shorthand for a sharp drop in prey numbers either induced or maintained by predators. Both wolves and predators suffer. The pits weaken the wolves, through hunger; on rare occasions a wolf may turn to cannibalism. But the wolves don't decline to the point where prey can quickly recover. The smaller numbers of wolves subsist on prey like beavers, muskrats, ducks, and hares—enabling the wolves to maintain low ungulate numbers. Both wolves and prey remain far below an area's carrying capacity, and these lows can persist for decades.

By this logic, the biologists are convinced that wolf control not only helps the prey rebound but actually is good for the wolves too. Reducing wolf numbers allows prey to rebound, and then wolf numbers can also climb back up.

The most intensive long-term research on wolves and prey has been conducted in Minnesota and on Isle Royale, Michigan. L. David Mech (his last name rhymes with "beach"), a preeminent expert on wolves, has conducted research in both places for over thirty years. A wolf biologist for the U.S. Fish and Wildlife Service for over twenty years, he has focused his studies on the wolves and white-tailed deer of northeastern Minnesota. Mech was one of the first to document the effects wolves can have on prey over time, although he generally takes a more moderate position on wolf control than do Stephenson and his colleagues in Alaska.

In the late 1960s, Mech's study area supported a large wolf population, and he estimates that on average a wolf kills about fifteen deer per year, though the number can go twice as high. Several severe winters with deep snow—just as in Alaska at the same time —made it easier for wolves to catch and kill deer. As wolf numbers increased, the white-tailed deer went into a steep decline. Soon the wolves were running out of prey.

In the early 1970s, wolves began to trespass into other packs' territories, searching for scarce deer. Wolf litter size shrank. Pups starved. Mech even began to find cannibalized wolves. By the winter of 1974–1975, wolf numbers had dropped as much as 55 percent.

For Mech, this research anticipated the findings in Alaska. His major criticism of the Alaska research, however, is that these lows do not necessitate wolf-control programs. He claims that the problem with the research in Alaska is that it has not followed the wolf-prey populations over a long enough period of time. Pointing to over three decades of studying the wolves and moose on Isle Royale, a 210-square-mile island in Lake Superior that may be the single best natural laboratory for wolf studies in the world, he argues that these predator lows are the down-curves of long-term cycles or fluctuations. In time, in natural circumstances, both wolves and prey rebound. Wolf control simply speeds up the cycles.

Mech's research has led him to rethink his view of the balance of nature. The problem with the concept of the balance of nature is that it creates too static an image of the relationships among the animals. He prefers to think of the relationship as a "dynamic equilibrium"—fluctuating over time as the result of built-in tensions and pressures.

Despite the studies of the Alaska Department of Fish and Game and L. David Mech, not all wolf biologists are convinced either that wolves depress prey numbers or that wolf control is justified. Gordon Haber, a biologist who studied wolves for many years at Denali National Park, is an intensely outspoken critic of the research and the killing of wolves. During the early seventies, when prey were crashing in Alaska, Haber did not see such an abrupt crash in wolves or moose in Denali National Park. This has led him to conclude that the wolf-moose system outside the park in the Tanana Flats had been stretched too thin by overhunting and thus was vulnerable to the catastrophes of severe winters.

To many critics of wolf control, it's not the wolves causing the crash in prey, but an artificial factor—overhunting. This is where biology crashes head-on with politics.

These critics turn to the decimation of Alaska's western Arctic caribou herd for further corroboration of their view. Once the largest herd in North America, it numbered 242,000 in 1972. With an annual harvest, mainly by native subsistence hunters, of about 25,000 caribou, the herd by 1976 had plunged to 75,000. The game limit was cut to 3000 a year, and the western Arctic caribou herd began to recover. By 1979, it had risen to 119,000—without formal wolf control.

The Alaska Department of Fish and Game estimates, however, that 200 wolves may have been killed by rural residents—mainly Eskimos—angry at hunting restrictions. The issue of native subsistence hunting, often on snowmobiles, complicates an already raw debate.

As the debate flares—in the press, in the courts, in people's living rooms—it easily turns into a fight between hunters and wolf advocates. As one environmentalist said to me, "It's our position that Fish and Game is managing for people. They're trying to meet hunting demand by killing wolves. But Alaska is not an exclusive club for hunters."

Though this seems to be how the political boundaries get drawn, most people, like Stephenson and Haber, stress that the lines of demarcation are not so clear. Stephenson says that such a portrayal of the situation is an oversimplification. According to Haber, well-monitored hunting is not inconsistent with healthy populations of wolves. I personally met several hunters and trappers in Alaska who were by no means fanatical wolf haters. Still, even Bob Stephenson admits that the question of hunting is an important factor: "People up here," he told me, "need their moose, want game in their freezers. In Alaska and Canada, we have to factor people into the equation. Twenty to thirty years in a natural cycle is a long time to wait for a hunter to get his moose."

There seems to be no question that the recent biological research has sharpened our view of the wolf as predator. We are less likely now to see the wolf as innocuous in its impacts on prey, aware as we have become of prey crashes and predator-prey cycles. Yet the disputes among biologists over factual matters—the exact causes of crashes, the census figures, the illegal hunting of wolves, and more—all show that the issue cannot be resolved by biology alone. We know we can give short-term benefits to hunters through wolf control. But we don't know the long-term implications of humans displacing wolves in the ongoing cycles between predator and prey.

It is clear that even a documented decline in prey numbers is not a necessary signal for wolf control. Biology here tells us what we *can* do. It does not tell us what we *should* do.

It is also clear to me that what is at stake in the issue of wolf control is not biological research—the ability of science to somehow show us what to do. What is at stake is how we will relate to wolves. What place will we offer them on our last frontier, as well as in the contiguous United States? This is a complicated question of politics and ethics. The attraction of trying to define wolf control as a biological problem is that we might be able to reduce our quandaries to empirical certainties, with an objective, external standard of right and wrong. We could absolve ourselves of the responsibility of choosing what to do, of deciding how to treat the wolves.

But the fact is we cannot absolve ourselves of responsibility. Science cannot tell us how to relate to wild animals, how to treat them, what place to make for them in our lives. Our relationships with animals cannot be reduced, or confined, to the comforting absolutes of a biologist's graph. In fact, the virtue of seeing our relationships with animals, both personal and cultural, as something we can choose, as a scenario created by us rather than determined for us, is that we are not bound by what has been. This view gives us the hope of freedom and change in our relations with animals.

What about wolf control? While it may be an oversimplification to see the conflict in terms of wolves versus hunters, I doubt there would be any interest in wolf control if humans were not part of the picture. Still, *which* biological arguments we find convincing—*for* wolf control or against it—are likely to be determined by how we answer one question: In a crunch, does a human being have more right to a moose than does a wolf?

III

This abstract and sophisticated research on wolves as predators is more than mere science. I want to consider it from another perspective, suggested by the curious intersection of scientific research and the politics of wolf control. If we learn to widen the focus, to increase the depth of field, when we view our interactions with endangered species, maybe we can begin to appreciate some of the subjective realities that shape even the most apparently objective view of a creature as controversial as the wolf.

I want to look briefly at the research on wolves not as science but as a cultural parable. In the context of history, the research on wolves as predators is not so much the formulation of knowledge as it is the reformulation of social values.

An eerie sense of déjà vu haunts the debate over wolves and wolf control in Alaska. There is a real question whether the killing of wolves in Alaska and Canada—sponsored by government agents, however sincere—is a repeat in another key of one of the environmental travesties of the lower forty-eight states. Is it a modulated twentieth-century attempt to do what the pioneers and colonists in our earlier centuries did with chilling thoroughness and an almost religious zeal in the contiguous United States? Then, the argument was to eliminate the wolf. Now, the argument is to reduce or "control" the wolf.

The historical evidence of the wolf's systematic persecution in the contiguous United States is overwhelming and is abundantly documented by Barry Lopez in *Of Wolves and Men*. In the colonial United States, the wolf must have once been very numerous, even allowing for the exaggerations in contemporary reports. In 1630, the Massachusetts Bay Company established the first bounty system in the New World—one penny per wolf. About then, William Wood wrote in *New England's Prospect*, "There is little hope of their [wolves'] utter destruction, the Countrey being so spacious, and they so numerous, travelling in Swamps by Kennells." According to Michael Wigglesworth, in "God's Controversy with New England" (1662), the world outside the colonial settlements of the Puritans was a place of devils, and his imagery evokes wolves:

> *A waste and howling wilderness*
> *Where none inhabited*
> *But hellish fiends and brutish men.*

This howling wilderness needed to be tamed. The colonists succeeded: By 1800, the wolf was eliminated from New England, lingering in eastern Canada in just a few pockets.

The history of wolves in New England comes as a sobering reminder of how quickly, how easily, how irrevocably we can destroy a species. As Peter Matthiessen's *Wildlife in America* shows, the early colonists looked upon a continent containing, to them, seemingly inexhaustible abundance. The teeming moose and wolves, waterfowl and fish staggered the imagination of early writers and explorers and left them certain the wealth could never be depleted. Matthiessen writes of "the Spaniards who extended the explorations of Columbus to the mainland . . . greeted by a variety of wildlife which will not be seen on the continent again." Yet the history of these animals—

their quick disappearance—is a slap to our confidence in a country's abundance. And the parallel with Alaska, and its wolves, needs to be drawn. Alaska is often described as a land so large that it is beyond our poor powers to destroy—yet that is exactly how the colonists felt about New England.

According to Barry Lopez, the worst carnage took place against wolves on the Great Plains. In their westward drive into the wilderness, the pioneers portrayed themselves as conquerors of a continent, bringing civilization into the darkness. This rationale extended to the clearing of forests and killing of animals. As Barry Lopez notes, Theodore Roosevelt, the great conservationist President, called the wolf "the beast of waste and desolation." The wolf was viewed as vermin, a scourge, a frightening predator, to be systematically persecuted. When white men destroyed the great herds of buffalo on which the plains wolf lived, the wolves turned to cattle. How serious the threat actually was is not clear, but the defense of cattle became the rationale for a full-scale campaign against the wolf. When the bounty system alone failed to kill all the wolves, the federal government underwrote a kind of "zoocide" using bounties, trapping, hunting, and poison. Between 1883 and 1918, Montana alone recorded 80,730 wolves killed, with bounty payments of $342,764, according to Barry Lopez.

By 1920, wolves were virtually eliminated throughout the west, their carcasses littering the path of Manifest Destiny.

Some of the frontier mentality also underlies the modern effort to control wolves. The man who led me to wolf dens, in fact, was an Alaskan frontiersman. Danny Grangaard not only works for Alaska Fish and Game on wolf-research projects but is also a triggerman in the wolf-control programs around Tok. He rides shotgun in planes and shoots packs of wolves in the winter snow. He is convinced that the wolf research ensures that wolves are being managed, not persecuted. As he said around the campfire one night, "No way are wolves gonna go tits up in Alaska."

Over twenty years ago, Danny left his ranch in South Dakota and settled in Tok, where he became a trapper. For two weeks I camped with him in mountains and marshes, warm sun and snow.

The whole time, he wore the same boots—cowboy boots. His idea of packing for a long outing? One extra pair of socks, a pair of leaky hip waders, and a carton of Marlboro cigarettes thrown into an old green Kelty pack. He is a thin man, with blue eyes that jump with intelligence.

I liked him—for his wildness, for his love of nature, and for his knowledge of wolves. I liked him especially when he tried to shock me. "Ya know what's a kick in the ass?" he asked me one evening, as he piled the campfire high with spruce branches. "Burnin' spruce trees. Ya get drunked up, go out in the snow at thirty below, and ya jest put a lighter on one o' the trees. Poof."

He said wolves were his favorite "critter," largely because they're so smart. When I asked him how he felt about shooting wolves in the control programs, he replied with ambivalence: "It's a bummer. But Mother Nature's a cruel sucker. Got to have some kind of management, cuz can't jest leave it to Mother Nature. She's tough."

Despite our political differences—he called me a "Sa-hara clubber"—I found Danny's personality engaging and clearly defined, and I tried to understand how he saw things. His perspective on wolf control fit with his view of nature as challenging and cruel. His description of the wolf as smart and tough was also a good description of himself, projected outward. His wolf was a reflection of himself, as is the case for all of us.

Which comes first, culture or science, is a tough issue. But it is hard *not* to see in all these views of the wolf the validation of cultural values. The interpretations of the wolf from earlier centuries—the frontier views—imposed moral values onto the wolf. Even the more recent, more objective views of this century reflect culture as well. The interpretations mirror particular historical moments, each with definite political implications when it comes to relating to wolves, to managing them.

How an objective science conditions our relations to wild animals came alive for me when I spent several days with L. David Mech on his research project in northeastern Minnesota. His book, *The Wolf*, remains definitive on nearly all aspects of wolf biology, and when I visited him he had radio collars on wolves in thirteen packs as part of his studies. Mech is highly articulate, very intelligent, and enormously stimulating to be around. He speaks with darting quickness, punctuated by active hands and flashing brown eyes. Over his many years working with wolves, he has acquired a forceful, authoritative tone.

Speaking with him about the disagreements among wolf biologists, I asked for his opinion on why there are so many points of view. He described his own position this way: "As a scientist, I subscribe to one method of determining truth in this world—the scientific one. I believe I'm the one objective scientist. Whatever happens, I just want to find out. I'm more of a scientist than a conservationist. I'm not attached. I try to minimize my personal interaction with the wolves." The power of this assertion, so direct and blunt, was refreshing. But I found myself pondering his words—"the one objective scientist"—for days.

Objectivity in science is a relatively easy target to knock off; the wonder is how tenaciously we continue to believe in it. As Susanne K. Langer writes in *Philosophy in a New Key*, "The formulation of experience which is contained within the intellectual horizon of an age and a society is determined, I believe, not so much by events and desires, as by the *basic concepts* at a people's disposal for analyzing and describing their adventures to their own understanding . . . these same experiences could be seen in many different lights, so the light in which they do appear depends on the genius of a people as well as the demands of the external occasion." (Italics in original.)

What I am suggesting is that objectivity is itself a cultural assumption, that it is one way among many of knowing an animal. Objectivity in studying nature requires a particular posture, a particular attitude toward animals. Objectivity, even rigorously applied, does not redeem us from the imposition of a point of view on nature. As Langer says, "The problem of observation is all but eclipsed by the problem of *meaning*. And the triumph of empiricism in science is jeopardized by the surprising truth that our *sense-data are primarily symbols*." (Italics in original.)

In other words, even the empirical observations we believe in as fact are *made*, not simply recorded. All phenomena, like crashes of prey or cycles of wolves and moose, have to be interpreted, given a history of cause and effect. Some ordering principle, some paradigm, has to be used to make sense out of experience, and the Western mind has turned largely to science.

Though it is true that the study of wolves, and other animals, as creatures has enabled us to see them better, it also imposes its own distance and limitations. Regardless of the knowledge it may produce, objectivity requires that we treat an animal as an object, not as a living creature that struggles, thinks, and feels—just as we do. It necessarily separates us from animals, yet we're surprised that we don't feel more connected. The blindness of empiricism is that it gives us too much faith in the illusions of our eyes. We come to believe that what we see is real. In the meanwhile, we have grown illiterate in the unconscious attachments to trees and rocks, and we are unaware of the deeper strategies and motives that shape our objective relationships to the world.

René Descartes, the seventeenth-century philosopher, was one of the founders of the scientific method. He is famous for bequeathing to posterity the concept that animals are to be viewed as mechanisms, *automata*—a paradigm we still labor under in trying to understand animals through general laws and statistical formulas. In Discourse 6 of the *Discourse on Method*, published in 1637 in France, he made clear the goals of this new philosophy of science,

> knowing the power and the effects of fire, water, air, the stars, the heavens and all the other bodies which surround us . . . , we might put them . . . to all the uses for which they are appropriate, and thereby make ourselves, as it were, masters and possessors of nature.

I have come to see wolf control as a case study, in extremis, of the kinds of effects we can expect in knowing animals objectively: distance from and power over nature. Managing wildlife is, after all, a contradiction in terms.

Some of these ideas were seething in my mind as I joined Mech in a small plane to fly over the winter snow of northern Minnesota to look for wild wolves. He was expert at finding the wolves. He used radiotelemetry, following the beeps in his earphones as well as the wolves' tracks in the snow. At the time, I had never before seen a wild wolf, but in one day Mech showed me seventeen of them, most of them sleeping on southern exposures to get as much warmth from the winter sun as possible. I was thrilled.

I have no doubt that the plane, flying over spruce bogs and frozen lakes, zeroing in on wolf packs I would never have seen from the ground, offers perhaps the best hope for keeping track of the wolves, for studying their populations, for gaining information useful in managing this endangered species. But the plane also seems to be a metaphor for a certain kind of relationship with wild animals. It offers the view from on high: commanding, expansive, and aloof. It is precisely here that the politics of wolf control and the politics of objective science intersect—both imply a kind of superiority to na-ture and a need to control it when it gets unruly. For all the pure motives of most of our wildlife managers—and I honor and respect their good intentions—wolf control nevertheless derives from the same world view that has enabled Americans to dominate nature wherever we have gone: Humans are superior to nature. If we no longer try to conquer or eliminate wolves, we at least try to control them.

This is the same premise that lies behind the conservationists' notion of stewardship in managing animals. The very idea of human stewardship, however, implies that we are in control of nature and responsible for it. We are flying above it, so to speak, striving for commanding views.

In the late afternoon, we flew over a single wolf crossing a frozen lake. The low winter sun stretched the wolf's shadow long and blue across the ice and snow. The wolf turned to look up at the plane, raised its snout into the air, opened its mouth, and seemed to howl.

We couldn't hear it.

For an instant I felt myself rise above the normal muddled course of my life. I wanted to hear a wolf howl, I wanted to see a wild adult wolf on the ground. I wanted, as it were, to see with my feet, connected to the earth.

It takes courage to admit that we are not absolved of our own particular points of view. That we all must look out upon the world from a particular time and place which involves us in distortions. That even our science is shaped by history. And that wolves are more than our poor projections, whether those projections derive from the moral view that exterminated wolves or the scientific view that controls them.

The pilot flew the plane in tight circles above the wolf so that we could watch it longer. The aircraft turned on one wing in a dizzying spiral. Looking down at the wolf, I tried to focus on it. I was happy for the view but aware, too, of the distance between it and me—of the distance we've put between ourselves and animals. We've rationalized nature, and made ourselves superior to it in the process.

I realized also that we cannot do to nature anything we are not already doing to ourselves.

IV

The three wolf pups shuffled out of their den, one by one, and plopped onto their haunches with a kind of sweet canine gracelessness. This was the same den I had earlier crawled into, staring so closely into the pups' wild eyes, transfixed in the glare of The Other. Only now I was hidden in the spruce trees, in a blind some distance away. I watched the pups for three days.

They tramped around the den site, schlepping their way through the days, bumbling into each other, and often tripping over their own oversized paws. They nuzzled each other with their little pug noses, bit faces and ears, and licked each other all over. In mock assaults, they even laid ambushes for one another, acting out playful rivalries as they pounced and sent each other rolling into the den, pencil-thin tails wagging furiously. They were irresistibly clumsy and, unable to stay away from each other, alive with body contact. For hours each day, they would sleep, lounging on top of each other in a rich puppy pile of bluish-gray fur.

As both Freudians and voyeurs know, looking can be an act of power. I have often had the vague sense that watching wildlife is a type of voyeurism. I know I like to watch. I was watching these pups at their den, grabbing glimpses from our distant blind, with that eerie and invulnerable sense of seeing but not being seen. There was something superior in my watching, as the pups played in the intimacy of their den, unaware of me. I had much the same pleasure that an audience has at a play, spectators who are always somehow superior to the characters on stage.

And like a play on the stage, the play of the pups was also a text to be read. Their play was anything but aimless. It was filled with meaning. Through play, these pups were learning to live in a pack. Wolves are intensely social animals, living in packs that are structured in rigid hierarchies. In the chain of power, each wolf has a defined place on a ladder of dominance and submission. In addition to developing distinct personalities, these wolf pups were also learning a complex vocabulary of gestures and postures for expressing their status relative to each other.

One of the pups, for example, was bigger than the others, easily identifiable by a bold white star on its chest. A leader, it was always the first one to emerge from the den in the morning. One morning, it poked out of the den but paused at the mouth for ten minutes, until the others appeared too. Then it padded off down a small trail into nearby grass. The other two followed. It chewed some grass with its head turned sideways, attacked an unsuspecting moose bone, disappeared into some spruce to the left of the den, and reappeared on a small ridge behind the den. Here, it sat down and scratched an ear, the quintessence of puppiness.

The smallest pup had returned to the mouth of the den, where it sat with its back to the pup with the white star. The starred pup suddenly began to get worked up with anticipation, spotting an opportunity. Pawing the ground, its rear end and little tail wagging in growing excitement, it looked like a dog eager for its owner to throw a ball. And then? A sudden spring, bowling the smaller, more timid littermate snout over tumbling tail into the den.

The triumphant pounce was only one of several ways the starred pup expressed its superiority over the others. Frequently, it would mount the other pups from the side, front paws on their backs, in a posture of dominance that biologists call "riding up." The other pups would respond with the classic pose of submission—head lowered, ears back, tail curled between hind legs. The starred pup would also assert itself by "standing across" another pup, which would curl on its back, stomach up, tail tucked between its legs.

Numbers in a wolf pack can vary greatly, from four to fifty wolves, but most packs contain six to fifteen. It's unclear how a pack comes into being, but most members are related. Normally, the dominant male and female, the "alpha pair," are the only ones to mate. Biologists believe most pups join their parents' pack.

Breeding starts as early as January through March, the farther north the later. Pups are born in April or May, weighing a pound. Blind and deaf at birth, they open their eyes at about fifteen days and are out of the den, exploring, by the fourth week. Sometime between the sixth and eighth week, the parents move the pups to a rendezvous site—a sort of lupine hub—where the whole pack helps raise them. The pups live at the rendezvous site while the pack goes off to hunt—in essence, all the adult wolves now become parents. The pups will beg for food from any member of the pack; the food is carried back to the pups in the adults' stomachs and regurgitated.

The deep bonds between members of a pack begin at birth and are reinforced throughout the wolves' lives. They teach a powerful lesson: The pack is family.

But a single text can always be read in multiple ways, and much of what I saw, watching the wolf pups over the days, was not natural history. In fact, I saw most of the natural history that I did only because I had learned it from reading or talking with wolf biologists. I could only partly view the pups as "things." They were other lives. I saw creatures that play, have babies, take care of them, live and die, think and feel, creatures that face the same problems life poses for us and that have all the dignity of doing what they must do to live. These pups had a secret and undescribed story—and this part of an animal, its *life*, is only apprehensible by wonder.

The powerful pull I feel toward animals comes from deep inside me. They stir feelings in me that lie below words, a primitive power. These wolf pups evoked deep echoes for me. I found myself responding to these pups not out of a sense of superiority or distance but in a more empathic and intimate way.

Often in the afternoon in the blind, I slipped in and out of naps as the wolves slept, and day naps for me swim with my most vivid dreams. In this unguarded state, I realized that the pups, playing and waiting for their parents to return—as I was waiting for the wolf parents to return—evoked in me memories of my childhood. Through dream and memory, I give shape to the world and make it live.

These wolves reminded me of a recurring scene from my childhood. Before a bitter divorce, my parents separated several times. My two sisters and I often entertained ourselves alone while my mom or dad was gone; we would eat candy corn, I remember, and watch *The Untouchables* or *The Roaring Twenties* on TV on Saturday nights. And just before I came to Alaska to watch the wolves, my wife and I had separated for the second time, and my two sons had asked me frequently when I would be coming back home. I felt this intense and irrational identification with the wolf pups waiting for their parents' return—a bridge contained in the image of the absent parent and the children on their own.

At some moments, I honestly could not have told you for sure where I ended and the wolves began. This is not mysticism—not oneness—though it is a breaking down of boundaries through feelings. At this level, when inner responds to outer, I feel most caught up in life, most in love with what the world holds.

Make no mistake, however. Moments like these at the wolf den are not pure release. I suddenly realized that all this longing I carry around with me—to see wild wolves, to crawl into their den—stems from a sense of loss. It reflects an emptiness, a sense of absence, that can probably never be filled.

Though the lyric form seems to dominate in most nature writing, I have to admit that I turn to nature not only to find the beautiful and healthy. My longings in nature lie close to pains and personal struggles. It is not easy to find positive images of the wolf in Western culture, but probably the most famous wolf story stresses the mother wolf as nurturer. According to the Roman historian Livy, Romulus and Remus were the illegitimate twin sons of Mars and a vestal virgin. The two boys were abandoned to die, but a she-wolf found them and suckled them so that they grew and became the founders of a great civilization.

I love this myth, for it holds the promise of a nurturing nature and it tells of the power that even lost children can draw from the wild wolf. It offers a concept I believe in not only for myself, when I feel most like a lost and motherless child, but also for all the endangered and threatened animals on our continent: that what has been lost under the domination of our civilization can be found and reclaimed.

I knew, watching those pups play, that what I really wanted to see was the mother wolf. I was willing to stay as long as it took to see her return to the den.

V

Charles Perrault, the French writer of fairy tales, wrote the classic and original version of "Little Red Riding Hood" in the late seventeenth century. Stressing the sexual danger of the wolf, Perrault depicts a dramatic confrontation between young girl and wolf in bed, and he does not have a hunter rescue Red Riding Hood (as do the Brothers Grimm). The end of the story comes swiftly, full of ominous threat and innuendo:

> "Grandmother dear, what big teeth you have!"
> "The better to eat you with!"
> With these words the wicked Wolf leapt upon Little Red Riding Hood and gobbled her up.

Perrault even provides a "moral" for French girls growing up in the age of Louis XIV. Don't listen to strangers, he warns urbanely, for even gentle wolves are of all creatures the "most dangerous." They follow "young maids in the streets, even to their homes."

In illustrating the tale, Gustavé Doré pictures Little Red Riding Hood in bed with the wolf. Doré picks the dramatic moment of recognition, when the virginal Little Red Riding Hood, wide-eyed and in bed with the wolf, seems suddenly to realize exactly what the wolf wants—the epiphany of sexuality.

This association between sex and wolves is no accident. In France, even today, people speak idiomatically of losing virginity as "seeing the eyes of the wolf."

For the most part, though, we've neutered nature. To adapt Lady Macbeth's great phrase, we have "unsexed" nature. We're not going to have some toothy, drooling wolf following us home from a trip through the woods or from our weekend camping trip to the national park. The nature photography in our magazines almost never acknowledges the sexual energy in nature. Instead, it seduces us with purely aesthetic or domestic images. Sometimes it presents the violent, which accords with the Darwinian view of nature as a struggle. Never, like the Greeks, do we show images of Pan, the god of nature, in his sexual guise—with his phallus huge and erect, too large to be hidden or controlled.

The Middle Ages taught us to moralize nature, and the Renaissance taught us to rationalize it. Still, when I turn to the medieval bestiaries—compiled before our distance from nature was as great as it is now—I can feel immediately the emotional concerns, the passions and vitality, and the energy of sex within the descriptions, only barely contained by medieval morality. A preoccupation with animals as images of lust, lasciviousness, and sexual infidelity is repeated in the descriptions of creature after creature. Those wanton beasts. The sexual and the emotional in nature, and in humans, had to be controlled. The entries on the wolf in the bestiaries are especially exemplary of these anxieties.

Derived from an early Greek text called *Physiologus* (The Naturalist), the bestiaries of the Middle Ages reached an apex in the twelfth and thirteenth centuries. They are compilations of natural history, Christian morals, and what we now call pseudoscience. As the scholar Florence McCullough describes them in her major study, *Medieval Latin and French Bestiaries*, "From the early centuries of our era through the Middle Ages the *Physiologus* and its later, expanded form, the bestiary, were among the most popular and important of Christian didactic works."

The bestiary tradition has, I think, been badly misinterpreted and maligned. From a post-Renaissance world view, most scholars look back upon the bestiaries and scour them for an emerging interest in objective natural history, the creatures themselves. There is some, but not very much, literal realism in the bestiaries, so the bestiaries are often treated with disdain or an embarrassed apology.

The bestiaries reveal a different consciousness. For the Middle Ages, the bestiaries *were* natural history, and they reveal the way one historical period made sense out of its relationship to the world. Often, contemporaries did lose sight of the creature itself and mistook the bestiaries' lore for literal fact. Monastic clerks in their cells offer encyclopedic descriptions of both beasts and monsters, with nothing to distinguish fact from fantasy—lions and lizards, hyenas and hydras, onagers and aspidochelones.

But the clerk writing a bestiary did not offer literal facts, because he was not concerned about them. The mind behind the entries in, say, T. H. White's *The Book of Beasts*, is not literal. It sees in different ways. White's book is a translation of a twelfth-century Latin manuscript now, according to White, in the Cambridge University Library. The imagination that wrote these entries sees the world as metaphorical. Nature is the text of God, and it should be read to reveal God's mysteries and the moral order that informs His creation. Call it a sacramental vision: The world of nature offers analogies to the human spirit. As T. H. White remarks, "The meaning of symbolism was so important to the medieval mind that St. Augustine stated in so many words that it did not matter whether certain animals existed; what did matter was what they meant." Beneath the didacticism and the moralizing, what the bestiaries give us is not pseudoscience but psychology encoded in the imagery of animals.

Amid a wealth of lore on wolves, the bestiaries stress the wolf's wildness, hunger, rapacity, and uncontrollable passions—the image of the demonic. In T. H. White's translation, the monastic writer summarizes the spirit of the wolf in the first chapter: "The word 'beasts' should properly be used about lions, leopards, tigers, wolves, foxes, dogs, monkeys and others which rage about with tooth and claw—with the exception of snakes. They are called Beasts because of the violence with which they rage, and are known as 'wild' (*ferus*) because they are accustomed to freedom by nature and are governed (*ferantur*) by their own wishes. They wander hither and thither, fancy free, and they go wherever they want to go."

Desire and freedom and fearful wildness—much of this is associated with emotions, especially rage. One bestiary text gives the etymological origin of the Greek for wolf, *lycos*, as a word meaning rage.

The wolf is especially revealed in its hunger, particularly its rapacity. Again in White's translation: "Moreover, the wolf is a rapacious beast, and hankering for gore. He keeps his strength in his chest and jaws: in his loins there is really very little." This is the wolf that stalked the lambfolds of medieval shepherds, one of the hounds of hell. In another etymology from the same bestiary, wolves "are called *lycos* in Greek on account of their bites, because they massacre anybody who passes by with a fury of greediness."

This rapacity was easily allegorized. Corrupt clergy were wolves; greedy barons were wolves. But in addition to representing greed or avarice, the wolf had religious significance for the medieval mind. Most especially, the wolf was like the devil. "The devil bears the similitude of a wolf . . . darkly prowling round the sheepfold of the faithful," says the bestiary. Jesus himself warned of the "ravening wolves" coming in sheep's clothing—false prophets preying on the unsuspecting. In John 10:12, Jesus called himself the shepherd protecting the sheep from the wolf.

Hunger and rapacity. The predatory wolf. For the medieval mind, sexual hunger was never far from the surface of the time's moralizing. The imagination of the medieval natural historians makes a profound psychological connection between lust and death, between passion and predation—as if the bestiarist realized that our lives in nature subject us to both. Says the bestiarist: What a wolf tramples does not live. Furthermore "Wolves are known for their rapacity, and for this reason we call prostitutes wolves, because they devastate the possessions of their lovers."

In classical Rome, in fact, *lupa* did mean prostitute as well as she-wolf, and the Lupercal temples became brothels. In some countries today, twilight at morning or night is still called "the hour of the wolf"—the hours when prostitutes and thieves are at work.

The association between wildness and sexuality in the wolf is made even more explicit. According to the bestiarist, it was reported that on the backside of the wolf "there is a small patch of aphrodisiac hair, which it plucks off with its teeth if it happens to be afraid of being caught, nor is this aphrodisiac hair for which people are trying to catch it of any use unless taken off alive." The aphrodisiac tuft of hair is only potent when taken from the wild, living wolf.

In the bestiary tradition, the wolf is seen in intensely physical ways. Its character resides in parts of its body—mouth, feet, tufts of hair. The body seems to provide images for psychological impulses. In animals, we can see images of our relationship to ourselves, projected outward. Read the bestiary backward, as it were, from nature to ourselves, and you find in the medieval lore revelations of some of the emotional dramas we experience in our relations with nature.

This is clearly shown in one of the most insightful pieces of medieval wolf lore. Much of the wolf's power is concentrated in its eyes. At night, says the bestiarist, the wolf's eyes shine like lamps, like a demonic glowing out of the dark. If the wolf sees a man first, it strikes him dumb "and triumphs over him like a victor over the voiceless." But if the wolf is seen first, it loses its own ferocity and cannot run.

Here, perhaps more clearly than in the other pieces of lore, is an allegory of control over the wild and dark in nature. Words and vision—two of our primary tools to steal the ferocity from the wolf. Our fear? That if the wolf sees us first, we lose our power of speech. We go mute and impotent. This, it seems to me, is a sort of primal fear of loss of identity, of the power of words to name things and give us control over them. Here is the fear that we may lose our power over nature, that we may lose ourselves in its power.

The real beast feared in the bestiaries is the beast within. In the struggle for control, it is easier to blame the wolf in nature than it is to face ourselves. The price of such control is steep—alienation not only from animals but from the vital and frightening energy within.

Beneath our words and abstractions, there is another realm of experience. As the bestiarist knew, this realm is visible in the demonic glow of a wolf's eyes. It is in that dumb and frightening silence beyond the reach of words, beyond the reach of morality, and beyond the reach of science.

I once saw an adult wolf very close up, a young one in a captive colony in St. Paul, Minnesota. He bounded up at me, onto my chest, like an overgrown dog, trying to lick my face. He threw himself on me, and I grabbed clumps of his fur in my hands. His yellow, depthless eyes were like bullets, fierce and wild even in captivity. Their intensity frightened me, disturbed me. When I leaned toward this wolf, one of his fangs gouged my forehead, drawing blood that trickled into my eye.

I don't want my life neutered and neat. I want to see those fierce yellow eyes and even to wrestle with the demonic and tear things open. I know the fear the medievals felt in the eyes and the fangs of the wolf. But I want also to hear the wolf inside, whose howls echo only dimly in the memory.

VI

After all my traveling in North America, looking for wolves, the central question remains: What place will we make for the wolf in our culture? In our lives?

The wolf, like other big predators, evokes powerful fears and intense passions. Like grizzly bears and mountain lions, the wolf poses major challenges: Will we preserve what range it has left in the United States? Will we reintroduce it into parts of its historic range in the contiguous United States? I had two very different encounters with adult wolves in the wild, and they suggest two poles in our experience of these predators—two postures we can take.

In late April 1985, I flew to the Eureka Weather Station, on

Ellesmere Island, because this was reported to be the best place in the High Arctic to see the arctic race of wolf, *Canis lupus arctos*. At 80° north latitude, it is a place that would appeal to any imagination in love with extremes. I went with a friend and photographer.

Located on the Fosheim Peninsula, the Canadian weather station where we stayed looked out on Slidre Fiord, a small arm off the Eureka Sound. In the afternoons, I could walk westward out over the ice on the fiord, to the edge of the sound, and see Axel Heiberg Island rolling into high hills in the cold, diaphanous light. The sun no longer set at night but rolled around the sky like a glowing ring, turning the snow and the arctic foxes into pastel shades of pink in its low and slanting light. But the cold was still strong. Icebergs lay trapped in the frozen fiord like huge hulks, images of arrested motion, as if the world had simply ground down to a halt.

On my walks around the island, I felt as though I had come to the edge of some austere absolutes—intense cold, a timeless and blinding light, immeasurable miles of blank snow. A white silence brooded over the ice and snow. There isn't that much snow on Ellesmere—in the polar desert, it gets about 15 inches a year, which doesn't melt until June. It was an immense void, swept clean and empty by the wind. I found myself thinking of arctic explorers, and of how some people will go such a long way to get a glimpse of the untouched and the ultimate.

It would be easy to romanticize my trip to Eureka, but that would be partly illusion. Other compelling images remain. The weather-station compound reminded me vaguely of a sprawling, self-contained cocoon: low buildings hunkered against the cold. The main building, where the six or so men lived while stationed at Eureka, was a vast dormitory complex with all the current accoutrements of comfort: wide-screen TV, VCR, library of movies (I saw *The Right Stuff* and *Gorky Park* while I was there), pool table, weight room. There was also a full-time cook and baker who kept us supplied with more food than we could eat—huge meals and a constant spread of cakes and cookies. I felt less that I was confronting the remote High Arctic than that I was insulated from it.

The first night, after dinner, one of the guys offered to give us a ride up to the dump. Arctic wolves hang out there, feeding off the garbage.

We scraped our leftovers into a bunch of plastic buckets, loaded them into the back of a pickup, and drove up the small hill between the bunkhouse, the airport runway, and the dump. We drove slowly, honking the horn like mad.

Looking to our right, out over the fiord, I spotted three white wolves a few hundred yards away. They moved superbly, their legs barely moving, in an ankle trot. Against the cool white world of ice and snow, they were like white ghosts, like white shadows even, drifting by.

A Russian proverb says, "The wolf is fed by its feet." Wolves travel all winter, and the urge to travel is probably related to the need to hunt. The wolf's chest is narrow and strong, its legs are long, and its paws are huge—all adaptations for a life on the move, even through deep winter snow. A wolf can cover a mile in about five minutes, effortlessly it seems, and keep the pace up for hours. It can cover up to 120 miles per day. When it spots quarry, the wolf prefers a short chase, a quick burst of about 100 yards, lips curled back, fangs out, snarling and barking.

When the three wolves heard us blasting away on the horn, they bolted toward us as if it were a hunt. All three of them caught up with us, fell in behind the truck, and followed us to the dump. I have vivid memories, poignant yet almost comical, of us driving, the horn blaring, and the wild wolves racing behind the truck like Pavlovian dogs after their dinner, ears pressed back against their heads, big snouts and heads pressed forward, long legs bounding under arching white backs.

At the garbage dump, we emptied the buckets and retreated. The wolves went straight for the food. Up close, through binoculars, their snouts and paws were smudged and greasy from eating at the dump, their white coats bleary with soot.

These were not the wild wolves I had come so far to see. Though they were wild, they seemed nearly tame. I found in these wolves an example of one way the Western mind accommodates all those disturbing impulses we find in the wolf—sex and violence, wildness and passion. We train it. We control it, as we "control" the wolves of Alaska. Both the literal and the symbolic wolf are left to live along the margins of our lives, feeding on our leftovers, as it were. It was a sad image, an image of loss.

But I have another image of wolves that counters this one and reveals a different way of relating to wolves. It comes from the den I watched with Danny Grangaard in the mountains around Tok, Alaska. The one I crawled into. The one where I watched the pups. The one where I got the urge to see the mother wolf.

After watching the den from the blind for some time, we decided we weren't going to see the parents come in while we were so close. Wild wolves are elusive. You have to be extremely clever, and lucky, to see one. Danny suggested we go up onto a bluff above the den, where we could camp and watch without the wolves' seeing, smelling, or hearing us. The bluff was very high, several hundred feet, and overlooked a gravel bar and several small ponds along Bone Creek. We made our camp under several big spruce trees. From our campfire, we could watch the bar below, though the actual den was out of sight in the trees.

We watched for four days, an almost constant vigil. I knew the scene below by heart. I could close my eyes and picture the creek entering the open gravel bar, rushing past and bending right in the distance between two very steep slopes that led straight up to mountain peaks. In my mind's eye, I knew just where the poplars were, at the base of the bluff; I could see the line of spring-fed ponds on the right, like milky-blue beads on a string, full of Dolly Varden rippling the surface as they fed in the evenings. Several times, I even imagined seeing an adult wolf step out of the bush and onto the clearing of the gravel bar. The watching only intensified my desire to see the mother wolf.

At three in the morning of the fourth night, a wolf woke me with its howling. I'd never heard anything like it before; its eerie notes were emphasized by the remote mountains and the smoldering ember of our fire. The howl was a manic moaning, rising and falling in pitch until it trailed off on a haunted, mournful note. I sat up in my sleeping bag and woke Danny.

A wolf howls to assert territory, to call meetings with other wolves, to entertain itself, or, as this one seemed to be doing, to announce its presence. Deep in the Alaska Range, the bluff was ringed with rugged peaks. Sky-piercing and heart-piercing, snow-veined relentless peaks. The howls and the mountains—all around, the beauty of the unattainable. In the slow alembic of a pale summer dawn, the snowcaps glowed in Botticellean blues and pinks. From the den below, the pups answered the adult's howl, a yapping chorus, more tremulous and piercingly high-pitched. The pups whined like a pack of coyotes. The adult answered them. The howls rolled down the mountainsides, echoing off the buttes and rock spires, the talus and cliffs, spilling out of distant recesses and inhuman reaches.

This was a true waking, wolf howls washing over me in waves, to something wild and summer-new.

Danny looked at me across the campfire. "If you're gonna see a wild wolf, this is your chance."

We got up, stood out on the bluff, and scanned the bar. Both of us breathed shallow, short heaves of expectation, alive to the imminent promise of satisfied hopes. And then the wolf appeared on the gravel bar in the grayish shadows below. She stepped from the greenish puff of aspens into the open, walking slowly, aimlessly, nonchalantly.

"That's the bitch," Danny whispered. He'd seen her several times before. She was all white, except for a steel-gray saddle on her haunches.

A few feet onto the gravel, she stopped and looked back over her shoulder, Bette Davis style, toward the den. She sniffed the air, padded on across the bar to the right, and wandered slowly toward our bluff. Her indolence was delicious to watch. Almost out of sight, she stopped, raised her snout, and howled again, a kind of hysteria in the voice so at odds with her casual pace.

And as if she were merely a specter, she vanished as phantomlike as she had appeared, into the timber at the base of the bluff.

Danny and I had been a rapt audience. A wild wolf. We can make room for them; we can let them live their own lives. The howl of the wolf speaks of regions in nature, and in ourselves, that we can never tame, never control. But we can learn to live with it. Occasionally, even if for only a few minutes, we can hear that demonic music in their howls, feel the ecstasy of those haunting echoes.

On the bluff, after she had gone, I let myself go in delirious surrender. I had gotten what I wanted, a look at the mother wolf. Sometimes we do get what we want; sometimes what is given is enough—like a mother/lover's breast in my mouth. Dancing like a dervish, lost in the euphoria of release, I was caught up in a sort of wolf madness, celebrating a wild wolf, feeling less like a spectator than, suddenly, a partner with this mother wolf in a dance I didn't for a minute understand.

TWO

THE FALL OF A SPARROW

I

In the figure of the wolf and its fate in North America, I find all the stuff of high drama. The wolf is a predator, endangered in the contiguous United States and persecuted still in Canada and Alaska. It is a creature with a long history in the human imagination. Rarely seen, the wolf can still rile primal passions and can still stir conflict and rage, whether the battle is over killing it up in the north or reintroducing it in the parks and wilds south of the Canadian-U.S. border. Its appeal to us goes deep, and makes me think of primitive epics, heroic conflicts—Beowulf fighting the diabolical Grendel in the fens, or Homeric passions unleashed in the war between Hector and the raging Achilles.

But for other animals, their destiny has a cruelty that can be suffered only by the inconsequential. For these animals, the story is simpler but more insidious—thousands of them vanish before we even know they exist. Creatures whose existence seems never to have mattered; creatures who, at least for us, seem never to have *been*. Creatures we don't quite know how to see, much less value.

Some are so far gone that by the time we notice them, it's too late. Others slide away, more quickly than we can believe, while some absurd human drama—the politics of extinction, call it—whirls around and about them, affecting them but absurdly irrelevant, in the end, to the creature itself.

The dusky seaside sparrow in Florida, called in Latin *Ammospiza maritima nigrescens*, was one of these animals.

An insignificant little creature, the dusky seaside sparrow took a fast plunge to its fate while we debated what to do about it. Its brief story reminds me of an existential tragicomedy, crazy enough at times to be funny, except that it is about life and death. I think of Samuel Beckett, of mundane characters living in a world utterly beyond their control and comprehension, where faint hopes keep crashing into frustration. Or of a modern story, tipping toward minimalism, where bruised characters bang into walls and slide to the floor, unable to fathom the senselessness of what is happening to them.

II

Try to get your mind around this surrealistic juxtaposition of images: The last dusky seaside sparrow died amid a monument to life as a comic cartoon, a monument to an America as innocent as Mickey Mouse or Snow White—he died in Disney World. A single dead sparrow in the midst of the American fantastic.

I visited Disney World in Orlando, Florida, in January 1986, when there were still two dusky seaside sparrows left, living in large cages on Discovery Island.

Both were males. Their fate was clearly fixed.

The sparrows were in the care of Herb Kale, a Florida Audubon Society biologist, and Charles Cook, curator of Disney World's Discovery Island Zoological Park. Herb had offered to take me to see them and even to let me enter their cages.

These birds represented the edge of extinction. I had a clear sense that this was a rare opportunity, something large and important, even vaguely historic. The other famous species of animals in the United States that had already gone extinct were very much in my mind: the Steller's sea cow, hunted to death by about 1768; the great auk, victim of overkill by 1844; the Labrador duck, hunted for feathers, gone by 1875; the passenger pigeon, killed in huge massacres, the last one dying in the Cincinnati Zoo in 1914; the Carolina parakeet, dying in a zoo in 1914; the San Clemente Bewick's wren, disappearing in the 1930s; the heath-hen, gone by about 1932; and the Smyrna seaside sparrow, a near relative of the dusky, vanishing sometime in this century, precise date unknown.

These two dusky seaside sparrows, though still alive, were as good as extinct. Extinction. A chance to look at it. Stare at it. Try to gauge it.

A few drops of rain splattered on the windshield as we pulled out of Herb's driveway. We turned onto the freeway, heading south through town toward Disney World.

"You're lucky you didn't get here any later," Herb was saying. "These birds could go at any time. We lost one just last fall."

In 1979, Herb Kale began a nearly desperate project to save the dusky seaside sparrow—a project beautiful and quixotic and doomed from the start. Herb is a highly respected ornithologist, with a list of publications that fills pages on his résumé. He had led a team through the marshes near Cape Canaveral, where the sparrow lived, intending to capture all the duskies they could find and breed them in captivity. All they could find, though, were six male sparrows. They caught five of them. No females—the last female dusky seaside sparrow was seen in 1975.

"We had hoped to find females," Herb said as we cruised down the freeway, doing only 45 mph in the Corolla. "But there were none. Everything had gone much quicker than we realized. After we caught the fourth bird, we went back to catch another, which we recognized from its band. We'd seen him on an earlier survey. We couldn't catch him. He must have been traveling miles, back and forth, singing. We figure he was looking for females, too, and he couldn't find any either. So we're pretty sure, if the males couldn't find any females, there weren't any, and we had the last of the males."

Instead of admitting the dusky was lost, Herb decided to try to "crossbreed" the five males with females of a very near relative, the Scott's seaside sparrow, whose scientific name is *Ammospiza maritima peninsulae*. Herb wanted to breed as close to a dusky seaside sparrow as he could get: Crossbreed and get a 50 percent dusky; then cross back to another dusky male and breed again; and so on, until he had a bird that would be up to 97 percent dusky, carrying the genes of an extinct race.

The five males were probably old when Herb caught them. One by one, they died. These last two were at least 11 years old in 1986, already aged for a sparrow. Both had various health problems, and one was infertile.

Herb is one of the most endearing men I have met in wildlife research and conservation: unaffected, genuinely kind, eccentric. In his fifties, he had a round face and brown hair and wore thick glasses. There was something about him that was sympathetic and even vulnerable. Perhaps it was expressed most obviously in his voice— a deep raspy wheeze, difficult to make out, the result of polyps on his vocal cords when he was a young man. He combined his ornithological knowledge with a kind of dogged persistence on his own course for the sparrows, blind somehow to the realities of the situation.

Herb Kale was virtually the only champion of the dusky seaside sparrow. Along with Charles Cook, he kept the fight alive after everyone else had abandoned the bird.

As he drove, explaining why he kept the faith, Herb got angry. "The U.S. Fish and Wildlife Service has used a bunch of arguments to claim the sparrow is already extinct, and so *they're* not doing anything. Everybody is abandoning them."

His indignation mounted. His speed on the freeway slowed. I realized cars were passing us like crazy. Then, suddenly, Herb slowed us way down. "Oh, my God," he said. "We just missed the exit on the freeway." He looked over at me with a confessional, slightly embarrassed grin: "I've ended up driving behind someone and following them to their home. I've driven right into their driveways."

But we recovered, drove through the high-rise hotels around Disney World, parked, and jumped on a boat to Discovery Island.

The idea of a small brown sparrow, increasingly an anachronism in its own life, spending its final years amid all the bright and exotic birds on the island was a wrenching anomaly. Pink flamingoes, scarlet ibises, green parrots, dusky sparrows.

We walked to the compound where the sparrows were housed: a long rectangle of two-by-fours, with a corrugated plastic roof and screened windows, open to the air. Pragmatic, low budget. We stepped through one door, scraped our shoes in a plastic bucket of green disinfectant—"to kill things," Herb said—and stepped in turn through a second door and another bucket of disinfectant. I looked down a concrete corridor in front of six cages.

Each cage was 10 feet by 10 feet. Sand in the cages simulated a beach, and clumps of cordgrass grew in tangles 3 feet tall. Into six cages, Herb had scattered seven sparrows carrying dusky seaside genes. Only two of them, of course, were full dusky seaside sparrows. The rest were the offspring of the crossbreeding project. One of the birds was a 50 percent dusky female, hatched in 1980. This half-dusky was mated with a pure dusky male (named "Yellow"), producing a 75 percent dusky female in 1983 and a 75 percent dusky male in 1984. The 75 percent female was then mated with another pure male (named "White"), producing an 87.5 percent dusky female, who was less than 2 years old at the time of my visit. The fifth partial dusky (25 percent) was a female produced in 1981 by a back-cross between a 50 percent dusky female and a pure Scott's seaside sparrow male.

"This is piss poor," Herb said. "We should have way more birds than this. I'd figured we'd have fifty to a hundred birds after this many years. Whadda we got? Five, plus two old ones. Half the eggs have been infertile, which happens with older birds."

Several of the tiny chicks that did hatch, about the size of a peanut, died.

In the second cage was the full-fledged dusky, White, named after the color of the legband he wore. The bird was only 6 inches long and had a stubby tail. Slate dark above, with a heavily streaked breast, he was otherwise nondescript. I hadn't expected the bird to be so dark, but that darkness, along with the heavy streaking, was exactly what made him a "dusky." It's what made these birds unique among seaside sparrows. The 75 percent and 87.5 percent duskies were indistinguishable from White, the pure dusky sparrow.

White was now infertile.

III

To me, White signified the small, lost, and insignificant life. Sparrows have long been a symbol of the lowly. In Matthew, 10:29–31, Jesus Christ asks, "Are not two sparrows sold for a farthing? And one of them shall not fall on the ground without your Father." Hamlet echoes the New Testament when, speaking to Horatio about his impending duel with Laertes, he says, "There's a special providence in the fall of a sparrow."

Over 500 species and subspecies of native animals and plants have become extinct in this country since the arrival of the colonists at Plymouth Rock: The Creator has had a lot to keep track of. This sparrow represented, to me, all the small and unknown plants and animals in North America that, lacking a constituency to defend them, become endangered and go extinct without a chance. Unlike, say, the wolf and other glamorous species that get the press and the money, sparrows bear the burden of our self-involved indifference and our blindness to the inconspicuous in nature.

In Florida alone many small creatures are vanishing: beach mouse, Lake Eustis pupfish, gray bat, Key deer, Eastern indigo snake, Stock Island tree snail, Schaus's swallowtail butterfly, Palm Springs cave crayfish, gopher tortoise, gopher frog, Harper's beauty, Miccosukee gooseberry, Key tree-cactus, and many more.

With its habitat confined to Merritt Island, within sight of Cape Canaveral, and the marshes along the St. John's River near Titusville, the dusky seaside sparrow vanished almost before anyone knew it was in trouble. Found only within Brevard County, Florida, the sparrow was once abundant, with a high of probably 2000 or more pairs in the first half of this century. But in the first comprehensive survey of the birds, conducted in the late sixties, researchers could find only about 1800; the population on Merritt Island was down to about thirty birds. One decade later, only six were left in the world.

The extinction of the dusky seaside sparrow began in the early sixties with the loss of its marshland habitat. Indirectly, too, it was the victim of our rush into space. The swampy marshes around Merritt Island were flooded to destroy the mosquitoes plaguing the development of the Kennedy Space Center and nearby Titusville. The high waters, created by impoundments, destroyed the marsh-grass habitat of the sparrow. Then, in order to make commuting more convenient for the space center workers, the Beeline Express-way was built between Orlando and Cape Canaveral, cutting right through the middle of one of the birds' marshes. Herb said he did his best to stall and modify that road. But his opposition just infu-riated most people—they were indignant that a little brown bird should hold up development of America's space program.

The marshes of the dusky seaside sparrow were also drained for housing projects and for rangeland for cattle. The cattle grazed in the cordgrass where the duskies bred, and because cattle like tender young grass, the marshes were frequently set afire during the winter dry season, destroying all but a handful of birds by the midseventies.

The dusky complicated matters: Like many endangered species, it had very particular habitat requirements. Not just any marsh would do. Duskies lived in cordgrass—a particular species, *Spartina bakerii*—at an elevation between 11 and 15 feet above sea level. These salt marshes need water in them, but if a marsh was too wet, the duskies couldn't feed.

Efforts were made to save the bird. The U.S. Fish and Wildlife Service tried to preserve habitat—the St. John's National Wildlife Refuge was established in 1971. According to Herb, however, the Fish and Wildlife Service dragged its feet in protecting the refuge from wildfires during winter and in plugging the ditches that were draining the waters away. The loss of the duskies was inexorable.

The dusky raised other issues, too, in the effort to marshal support on its behalf. Since the crossbreeding with the pure males has produced partial dusky offspring, the question arises of whether Herb has been saving something other than duskies: a diluted hybrid. Not to Herb. He has been saving dusky genes, which are real to him. When he spoke about the partial duskies, it was as if he saw them as a biological entity, a container for genes. He saw through the individual bird, to something that was invisible to me.

Also, the dusky is not a distinct species. It is a subspecies. This is a critical fact, for in our society a great deal hinges upon a name. Until 1973, the dusky seaside sparrow was considered a separate species, and that qualified it for financial support from the U.S. Fish and Wildlife Service. But then its taxonomic status changed. Though it was once much more common for ornithologists to be "splitters," dividing one species into several different species, it is now the fashion for them to be "lumpers," combining species. When the dusky became a mere subspecies, it lost most of its public support. The money from the feds dried up—subspecies are relatively low-priority creatures—and Herb had to find private support for his crossbreeding effort. He generated it through Disney World. More distressing, since birdwatching is a sport built upon finding distinct species, birdwatchers abandoned the dusky seaside sparrow. The small bird lost even this constituency. In a real sense, the dusky was a victim of the categories humans impose on the world and of the arbitrary nature of a name.

IV

Herb and I walked down the corridor and stopped in front of the fourth cage, Orange's cage. The second surviving dusky, Orange was blind in one eye. In the cage with him was his mate, a 75 percent dusky. With all the hyperactive, flitting, stuttering energy I associate with sparrows, Orange burst from his cover in the grass and flew to the screen; clinging to the mesh sideways, he looked about.

Herb offered to leave me alone in the cage with Orange. I stepped carefully inside and Herb left. A Disney World train whistled in the background.

In the cage: an intimate moment. I wanted to feel it, to be sure I was fully there.

I thought of my favorite Roman poet, Catullus, and all his Latin passion. In one lyric poem he envies a pet sparrow's intimacy with his love, Lesbia. And in another, dedicated to the sparrow itself after its death, he mourns the bird with an exaggerated and moving solemnity. "*Passer, deliciae meae puelae,*" he writes. "Sparrow of my sweet delicious darling."

And I thought of John Keats. In one of his letters, he uses the sparrow to describe one of the great human talents, one of the powers of the poet, which he calls "negative capability." It is the power of the imagination to enter into another being, to participate in the immediate experience of nature:

> The setting sun will always set me to rights—or if a sparrow come before my window I take part in its existence and pick about the gravel.

But me in the cage? An epiphany? Not even close.

I felt nothing—just an immense failure of imagination. Extinction is supposed to be a biological apocalypse, a catastrophe. Yet no bridges collapsed. There were no explosions or revelations, no sudden and surreal visions of the end.

Just emptiness on a gray and rainy day. The sterility of sand and everyday life. Both the sparrows, Orange and his mate, kept their sparrow-distance, staying hidden within the cordgrass, sparrow-shy and sparrow-secretive. Darting through the grass, Orange occasionally popped into the open to look around. He hopped up onto the wire screen, his plump body exposed, small feet grasping the mesh, head cocking about. He flitted back down to the sand and scooted to the other end of the cage, always keeping the grass between us.

Death in Disney World. It was as if nothing out of the ordinary were happening. Life as usual on the way to oblivion. All I felt was an awful, lonely gulf between the sparrow and me.

I got out of the cage when Herb returned.

V

Orange was the last dusky seaside sparrow. He died in that cage on a Tuesday morning. Though he was not, at the very end of his life, as active and lively as he had been in earlier years, he showed no signs in the weeks before his death that he was dying. He maintained his weight. Plans had been made to take him to a hospital if he showed indications of weakness. At 7:30 a.m. on June 16, 1987, he was seen alive in his cage. At nine that same morning, an attendant came back and found Orange dead, lying in his dish.

I'm not sure there are any lessons to be learned from the extinction of the dusky seaside sparrow. Even nature itself seemed to conspire against the little sparrow. On March 27, 1989, a windstorm knocked a palm frond through the roof of the compound. One of the remaining Scott's and dusky crossbred sparrows died. The rest vanished. Searches turned up nothing. As of spring 1989, even the dusky seaside sparrow's genes are extinct. One decade after Herb had begun a heroic and hopeless effort on behalf of the bird, the project ended in failure. Nothing of the dusky remains to mitigate our sense of guilt, impotence, and loss. The bird is irrevocably gone.

But when I went into Orange's cage, the sparrow's extinction was still in the future. Later that evening, Herb and I had dinner with other Florida Audubon people and then headed for his home just after midnight. A gloomy storm had clotted the sky during the day and dropped heavy rain—the same storm that delayed the launch of the *Challenger* space shuttle. Herb told me that he could no longer go back to the marshes around the St. John's River where the duskies used to be—"Just too sad." He revealed, too, that he kept working on the dusky seaside sparrow partly to be a constant goad to the U.S. Fish and Wildlife Service, a reminder to them that they had abandoned this bird.

He also told me that he had written an obituary for dusky seaside sparrows, though he wouldn't show it to me. This was an intimate confession for him, since he tried to stay hardheaded in his view of the birds, seeing them largely from a biological perspective.

Just north of Orlando, about 3 miles from the exit to Herb's house, the Corolla started sputtering. We coasted to a stop, out of gas. I looked over at Herb, who was unperturbed. Herb apparently runs out of gas often. He can be so absentminded that he forgets to check the gas gauge. For such emergencies, he keeps blankets in the car, and a gas can—which, unfortunately, was also empty. This is the same guy who, earlier in the day, missed an exit, and accidentally follows strangers home, the guy who lives in a world of genes and crossbreeding and impossible causes.

Except for a few passing cars, the freeway was dark and empty, shiny in the lamplight after the day's rain. We decided to take the gas can and search for a gas station.

Looking back on that night, I see the two of us in that absurd out-of-gas Corolla. And I think of society's ambition to reach space, its compulsion to build freeways. I compare these with the loss of the dusky seaside sparrow—with the wounds we've made as we've chased our ambitions. In his futile efforts to save a small bird, in his hopeless attempt to mend some of the damage that we've done, Herb himself is much of the meaning of the dusky seaside sparrow for me.

Together we hiked down an exit ramp, late at night, looking for gas.

THREE

CARRION FOR CONDORS

I

I wasn't supposed to see what I saw. It had taken a month of patient persistence and all the subtleties of deference to get access to these condors, to be in this place at this time. It made me a witness—not to a nicely managed public-relations triumph but to a secret no one wanted to own.

Across the canyon, a condor in a pine tree was acting strange. For years, this condor had been intensely studied by the biologists I was with: Her life and her habits had been scrutinized, her nest closely watched, her mating studied, her eggs even taken to zoos to hatch, her babies raised in captivity. The most intimate moments of her life were known. But now she sat there like a stranger.

She was AC-3—"Adult Condor Three." Her mate, AC-2, sat just below her in the same pine. Of the six California condors (called in Latin *Gymnogyps californicus*) left in the wild, they were the last successfully breeding pair. They were the rarest birds in North America.

Now, in December 1985, something was wrong with AC-3. Except for very short flights, the two condors had not left this roost for over a week. That was not like this pair. Their habits were well known, since all the wild condors at this point wore radio transmitters and were followed daily. Normally, AC-3 and AC-2 foraged in the arid grasslands and on the ranches during the day and flew back to their remote roost in the mountains at night. Something was definitely unusual—maybe AC-3 was very sick. Though she looked the same, her behavior made her unrecognizable to the people who knew her. AC-3 was a stranger with a frightening secret.

Her long-time mate, AC-2, stayed with her all day, stoic, unmoving, patient, and loyal—if such qualities can be applied to condors. For hours they sat, neither of them budging from their melancholy perch. Condors are the biggest birds in North America, amazingly big. And black. In flight, with their wings nearly 10 feet tip-to-tip, they have sometimes been mistaken for small planes. They are also vultures. Hunkered on their roost, these two condors sucked me into their spell of macabre and dubious majesty. It felt as if it was watching death waiting to happen.

Occasionally, AC-3 would shake out her wings, immense and black with white triangles in the lining—the classic identification of a condor. She reminded me of a priest folding himself into broad, black vestments. Her perch was about half a mile from us, across Bittercreek Gorge, a steep ravine pleated with ridges and alive with the splendid greens of pines and live oaks. The morning sun spread in slow saffrons over the hills like color coming into lovely cheeks. Most of the gorge was still in shadows, and in the recesses mist lingered, deepening in the distances, brushing the canyon with a smokey, *sfumato* mysteriousness. Perched in the top of a pine, AC-3 was just above the shadows, and the slanting sun glanced off her ebony body with a dazzling iridescence, the liquefaction of a solid black bird.

Yet for all the magnificence of her presence, I wouldn't want to cloak her too much in the romanticism of the sacred. If there was mystery in AC-3, it was the horrific mystery of death. Her immense head was featherless—well suited for rummaging in a rotting carcass of deer or cow without getting fouled by guts, grizzle, and gore. In the honest brightness of the morning sun, her bald head shined pink, shading to dull oranges and scaly flesh tones, almost like the face of a person who'd held her breath too long, all red and puffy. Her beak was huge and powerful, for grabbing and ripping meat. With her beak closed, she had a sort of pursy, senile smirk on her face.

Was that the leer of grinning death? The leer of her own death?

There was in AC-3 and her uncertain condition a summary of her entire species: Something was wrong, and matters were getting worse, but nobody knew what the problem was, much less what to do to save her or the others.

The story of the California condor in the twentieth century has been a story of death and loss happening right before our eyes, as we watch helplessly. Estimates of the condor population in the past vary widely and, since the bird loves to fly over huge ranges, are necessarily very rough. The best estimates nevertheless trace a steady, ineluctable decline. Major losses seemed to come between 1880 and 1920, caused largely by shooting and egg collecting. Before then, the bird soared in skies as far north as Washington and even British Columbia—Lewis and Clark reported condors on their journey down the Columbia River. (They also shot the condors they could catch up to.) Between 1920 and 1950, the condor's range had shrunk to southern California. The best estimate for that time, by Carl Koford, put their population at about sixty—though that number is now generally believed to have been far too low; the consensus is that there were probably about 150 in the 1940s. They were being studied by then, and land was being set aside for preserves. In 1968, the estimate was fifty to sixty condors. By 1975, the number had dropped to forty-five. The population declined suddenly to about twenty in 1980, when a major effort was galvanized to save them through the formation of the Condor Research Center.

Still, matters worsened. In 1984, despite a staff of twelve full-time people, biologists for the Condor Research Center knew of only fifteen condors in the wild, with five breeding pairs. But in some ways, the future for the species looked brighter than it had for decades, since a captive flock of eighteen condors had been created, amid intense controversy, in two zoos—the Los Angeles Zoo and the San Diego Wild Animal Park.

Then came a disastrous winter. By March 1985, biologists realized that six of the wild condors were missing. In the space of one winter, 40 percent of the wild population had vanished, and only one condor carcass had been found to suggest why (high levels of zinc and lead in its blood). Worse, in this latest group of losses, four of the last five pairs of breeding condors either died or were broken up, leaving only AC-3 and AC-2.

Of the nine condors remaining in the wild, three more were captured and placed in the zoos. The other six stayed in the wild, and their future was the subject of bitter national debate—the stuff of headlines in *USA Today* and features on the evening news with Dan Rather. The controversy was whether we should leave these six in the wild, try to protect them from whatever was killing them, and try to preserve habitat for future condors if the species were ever restored or whether we should capture all the remaining birds, put them in the zoos, and pin all hope for the species on a program of captive breeding.

On Tuesday, December 17, 1985—the day that I was in southern California observing AC-3 in the field—the U.S. Fish and Wildlife Service announced that it would capture the last six California condors in the wild. Even as I watched AC-3, wondering what was wrong with her, I knew I was seeing the end of the condors in the wild.

And in the dramatic crisis over condors and what to do for them, I knew—I could feel it everywhere on the condor project— that this was a major moment in the history of endangered species in North America. To give up on saving the California condor in the wild, at least for now, marked a significant change in our understanding of the problem of endangered species and how to address it. For behind the debate over wild condors versus captive condors —so passionate and disturbing—I found another question, much simpler and more telling in the long run: What had gone wrong with the entire effort to save California condors?

Despite all the talk and publicity, all the newspaper reporters and TV crews, all the press conferences and official positions, there remained this enigma. It was a secret we seemed to be keeping even from ourselves as we talked. Why—after millions of dollars, after half a century of conservation efforts, after at that time five years of desperate biological triage for the species—why had we been reduced to removing the condors from the wild?

The answer to this question is deeper than mere biology. Deeper than any specific causes of mortality in the condors. Deeper than shooting or collecting or lead poisoning or habitat loss. The answer to this one lies in us. It is a question of cultural authority over endangered species, a question of power and impotence.

In the California condor, we are on the verge of an extinction we have not been able to prevent. Despite our efforts, we had been taken to an extreme reality: six left, and counting—the edge of the void, the madness of frustration over the pervasiveness of death. So few birds left, in a species that had come to symbolize the efforts of an entire culture on behalf of all endangered species. Small wonder feelings ran so high, recriminations so fierce. Over the California condor, emotions worked on the edge of desperation for years, the terms growing increasingly more absolute for death and extinction, more uncompromising. The project finally broke in a paroxysm of finality and futility—the decision to "bring the birds in."

This decision was not a giving up. It seemed, and still does seem, the last and best hope for the condors. Yet, despite the optimism that was attempted in press conferences and news releases, having to capture the condors was anything but a triumph. It was impossible not to feel, especially as I looked at AC-3, that something had happened we could not control. Some impotence in us was exposed, leaving us with that bleak sense that "the Queen is dead."

The sad destiny of the condor seemed to be acting itself out again in the life of AC-3. Everyone was worried about her. Between the deaths and the decisions, her life had grown even more precious, as life always does when tragedy grows more imminent.

The three of us on the ranch—one biologist, one veterinarian, and I—kept trying to invent reasons for her lethargy, barely admitting, if not outright avoiding, what we most feared: AC-3 was sick and in serious trouble.

We studied her for clues to her secret. Maybe her gullet was swollen. When they find carrion, condors often gorge themselves, filling up and then becoming sedentary. Maybe AC-3 had just eaten—a cow carcass was nearby. Her pink neck bulged noticeably—perhaps the sign of a full crop. She turned her head behind, slightly cocked, and preened under her wings, lifting them slightly, rubbing and rolling her head over her back like a duck. The black ruff feathers, encircling the base of her neck like a feather boa, seemed to flutter and stick out. We hoped it might be a sign that she was full of food.

Yet we also knew that more might be wrong. On November 1, AC-3 had been caught so that her radio could be changed. Blood samples were also taken, and were analyzed after her release. Tests revealed high levels of lead in her system—the same problem two other condors were known to have died of in the recent past. The birds pick up the lead from bullets in deer carcasses, and it poisons them slowly, insidiously. Was this what was wrong with AC-3?

Through the languors of a long, warm winter afternoon, I watched AC-3 and her mate, afraid that she would either die or be captured. I began to feel as though I were the vulture, watching the condor, waiting for it to go. The real vulture was now the victim. AC-3 became for me an image of the soft and vulnerable underbelly of a dying condor program. In a quick reversal, a spinning paradox, all our best efforts seemed like little more than gestures, so much carrion for condors.

About three in the afternoon, a third condor flew in and circled near the roosting pair until all three took off down the gorge. Sailing just over the trees, they flew beneath our position on the ridge, black wings spread over black shadows. The giant birds labored to get some lift under their wings, flapping them several times and then gliding short spurts in heavy, struggling flight.

The air in the gorge must have been too cool, too dead. The condors circled wide and headed back up the gorge. On the upbeat of their wings, the white linings showed, long triangles nearly the length of the wings, flashing like knife blades in the sun. The huge birds passed over the carcass on the plateau just below us; about half a mile up the canyon, they dropped ponderously onto a big pine on the far side.

Through a scope, I looked at the condors and then down at the cow carcass. It lay near some bushes, flat on its side, a patch of December snow refrigerating it in the shade. Half the black and white skin had rotted off, exposing the ribs. Extruding from the rear was a dead calf. Cow and calf had died in labor. During the day, the warm winds had brought us whiffs of sickening perfume from the rot, the sweet putrescence of death-in-life.

II

One outspoken local man I talked to was disgusted by condors. He called them, succinctly, "dirty black birds, flying garbage cans." Powerful words, powerful point of view—he spoke bluntly. As we talked, we shared not attitudes but images of the condor or, really, of vultures.

We all carry within us a repository of vulture images, most of them associated with death, decay, and physical degeneration. These associations usually cause us to recoil from vultures in disgust, inheriting as we do some of Hamlet's hatred of his "too sullied flesh." Even so, inclined as I am to take a subversive view of things, I find something liberating in liking the condor. I am drawn to it.

What I want to be liberated from is centuries of cultural authority that has decreed the vulture a vile bird because it feeds on rotting flesh. Emotional responses to the vulture, as to other animals, are so strong that we are blind to their arbitrariness. And this is what ultimately victimizes the condor, the wolf, the sparrow, and other animals: our attitudes, which we mistake for reality. What we don't see—and what is very hard to see—is that we choose the facts we believe. I feel tyrannized by beliefs I don't understand, can't determine the origins of, or, worst of all, can't even see as they shape my world.

I can feel inside me attitudes toward condors that go way beyond my making; coming from deep within our culture, they hold me thrall even though I sense them and yearn for freedom. Unable to talk myself into a new way of seeing, I try to recognize the origins of my attitudes in cultural images and beliefs, and to see what new ways of responding to nature emerge when I get to some basic image or undifferentiated core.

Take condors. The loathing of vultures stinks of deep cultural roots, so ingrained that our fears seem instinctive. The association between vultures and carrion has festered in the Western imagination since at least the Old Testament. Leviticus 11 includes the "carrion eagle" in a long list of birds that are an "abomination among fowls," and the gloss in the King James Version identifies the carrion eagle as the vulture. What lies under this categorizing of the vulture seems to be a priestly, protopuritanical anxiety about what is clean and unclean in nature.

The medieval bestiaries offer lore about vultures that is an easily understandable combination of immediate, frightening experience and inference. The most gruesome legends tell of the horrors of medieval war. It was thought that vultures had a desire to follow armies for the inevitable carrion and that they were even able to smell carrion from as far away as a three-day journey. It was thought that one could foretell the number of dead from a battle by the length of the column of black birds trailing the armies. When a vulture found a human corpse, it would go straight for the soft, vulnerable, and most precious organ: It would eat the eyes first and then suck the brain out through the orbits. Because they fly at such great heights, vultures were also thought to see cadavers at fantastic distances, even beyond the seas, so keen was their eyesight.

There is something nauseating and gripping about these associations in the medieval mind. In the pictures that accompany some bestiary entries, the vultures tear at animal flesh, hold human arms or legs in their beaks, or place large talons on the shoulder of a human corpse. There is an almost hallucinatory, nightmarish savagery to these images of life felt at the bone: It is not just the general concept of death that the vulture suggests, but human death—and, beyond death, dismemberment and disintegration. Reflected in the vulture is our primal fear of being ripped apart, of being fed upon.

The same revulsion informs Oliver Goldsmith's *History of Animated Nature*, written in the eighteenth century. He describes the vultures of Brazil, for example:

The sloth, the filth, and the voraciousness, of these birds, almost exceed credibility. In the Brazils, where they are found in great abundance, when they light upon a carcass, which they have liberty to tear at their ease, they so gorge themselves that they are unable to fly, but keep hopping along the ground when pursued. . . . But they soon get rid of their burden, for they have a method of vomiting up what they have eaten.

By the neoclassical age, the medieval horrors seen in vultures had been tamed. The associations had diminished to a preoccupation with an engorged vulture, reflecting an Augustan concern with manners and decorum.

In Western culture, the distrust of nature finds expression in the image of the vulture as a carrion creature. Implicit in this concept is the ultimate betrayal we all face both in nature and from nature: our own deaths. The vulture, eating, reminds us of what in nature will eat us.

Our century has inherited these anxieties and turned them inward. We have begun feeding upon ourselves—a self-consuming culture at the end of its days. Literally, we have devoured nature and are close to destroying ourselves with overpopulation and nuclear weapons; but psychically, too, we are feeding on ourselves, as a sonnet by Miquel de Unamuno suggests. A member of the "Generation of 1898" in Spain, he writes of the "voracious vulture":

Este buitre voraz de ceño torvo
que me devora las entrañas fiero
y ea mi único constante compañero
labra mis penas con su pico corvo.

El día en que le toque el postrer sorbo
apurar de mi negra sangre, quiero
que me dejéis con él solo y señero
un momento, sin nadie como estorbo.

Pues quiero, triunfo haciendo mi agonía,
mientras él mi último despojo traga
soprender en sus ojos la sombría

mirada al ver la suerte que le amaga
sin esta presa en que satisfacía
el hambre atroz que nunca se le apaga.

In translation:

This voracious vulture with a grim countenance,
who fiercely devours my entrails
and is my only constant companion,
carves my sorrows with his hooked beak.

The day that he takes the last gulp
of my black blood, I want
you to leave me alone with him
a moment, with no one to bother us.

Since I want to make a triumph of my dying,
as he swallows my last remains,
surprising in his eyes the gloomy

look when he sees the fate that awaits him
with this prey with which he diminished
his atrocious hunger that is never satisfied.

The vulture within has become the existential image of inner torment, and death becomes its own kind of gloating, personal triumph over the horrors of life.

Other possibilities for the vulture can be seen in cultures with radically different views of death. Writing in the late fourth century, Horapollo put together a synthesis of Greek science and oriental religion in a piece of natural history called the *Hieroglyphics*. He shows a system of correspondences that link the lives of animals and humans. To the Egyptians, he says, the vulture was not dirty. In fact, it was a great symbol of purification, compassion, and maternity. Mother goddesses had heads of vultures or other vulture attributes. The goddess of eternity, Maut, was associated as well with the vulture.

Since the world of death is the real world, the world of the afterlife we will enter as mummies in our tombs, the vulture could help us on our passage. The vulture could nurture us. Horapollo writes of the Egyptians:

> When they mean a mother, or sight, or boundaries, or fore-knowledge . . . they draw a vulture. A mother, since there is no male in this species of animal. And they are born in this way: when the vulture hungers after conception, she opens her sexual organ to the North Wind and is covered by him for five days. During this period, she takes neither food nor drink, yearning for child-bearing. . . . But when the vultures are impregnated by the wind, their eggs are fertile.

Patristic writers of the Catholic church took this lore for fact—having, I suppose, no reason to doubt it. St. Ambrose, writing in the fourth century, saw in nature literal truths that confirmed God's message. Nature was the script of God's text. In the vulture he saw a corroboration of the virgin birth of Christ:

> Indeed vultures are denied to indulge in coition . . . and thus without any mate they conceive by seed and generate without conjunction. . . . What say those who are accustomed to smile at our mysteries when they hear that a virgin may generate . . . ? Is that thing thought impossible in the Mother of God which is not denied to be possible in vultures? A bird bears without a mate and none confutes it, and because Mary bore when betrothed they question her chastity. Do we not perceive that the Lord sent beforehand many examples from nature itself by which incarnations he proved the virtue of the suspected one.

In Horapollo and St. Ambrose, we have not just another story about vultures, but "facts" based on entirely different structures of belief.

I find all these different representations of vultures humbling. Is it not likely that we, too, are each bound by our individual points of view, as invisible to us as those of these great writers were to them?

I know that, in addition to the desire nature can evoke in me,

I also feel a genuine distrust of nature, a sense of a great disturbing presence out there. I fear its strangeness. However much I seem to have located my life in nature, I still feel the presence of fearful and insistent figures, strange animals, vultures trailing me. How much of this is cultural and historical rather than merely a fact of life? Like most people in our culture, I can celebrate birth but am much more hesitant in my response to death. In our medicine, our ethics, our economics, we promote life—the great positive—and try to make it more abundant by defeating death. We bury death under the wealth that proliferates all around us, so visibly expressed in the high-speed images flashing across our video screens, images we see but never comprehend.

I stare at AC-3, a condor perched above a rotting cow, a condor that I'm afraid is dying, and I wonder what secrets she knows about death and decay and sickness. Even now, I shudder as I recall the condor in the tree and the pile of guts on the ground. It is not just physical revulsion I feel but a shudder of distant recognition: I want to manage and control my own death. It seems that endangered species in general have taken on a sort of social function for us: As we make life more abundant for humans, we displace death onto more and more animals. We have institutionalized them— made them rare, conferred status on them as endangered and wor- thy, and excluded them from our lives. Managing them along the periphery of our culture, they are like the vultures that hover on the edges of our lives. Endangered species are one way that we have made room for death-in-life.

At base, I fear that I am little more than carrion for condors. For all my assumed dignity, I am really a dying body, a slowly rotting carcass, meat for worms. I feed on my own guts and I am devoured by my own diseases—consumed and nourished at once, dying even as I'm growing.

III

On Easter Sunday, 1987, the last of the California condors in the wild was captured and transferred to a zoo. He was AC-9.

Though this seemed to bring to a close a long conflict, a scandal really, over what to do with the condors, it was more like the end of a denouement. It was the last of several spasms following the climax that had come in December 1985, when the U.S. Fish and Wildlife Service announced that all the condors were to be captured. The decision to bring in the condors—itself a major change in policy for the Fish and Wildlife Service—was a watershed, validating a new approach to the growing problem of severely endangered species like the condor.

In the ecodrama over condors, the bird became a social test case, apart from the crucial question of its future. And though it sounds rather grandiose to talk this way, it is nevertheless true that we entered a new era. However confused the debate became—between conservationists and zoos, between biologists and bureaucrats, between (as one person said) ecosystems and egosystems—at base was the question of cultural authority over the future of endangered species. Limiting the issue to biology would be like thinking that the tragedy of *Othello* lay in Desdemona's handkerchief rather than in all that that handkerchief represented in Othello's mind or Shakespeare's England.

It was a struggle for power within the biological community, within the community of "bio-power," to use Michel Foucault's terminology—the whole disparate community of environmentalists and academics and field biologists and governmental institutions that have more or less taken on the task of saving our wildlife. And in the politics of this agon over condors, in the underlying terms of the battle, we can understand the larger social issues at stake. The issue over condors is implicit as well in the battles over wolf control: Even the "biology" advocated by parties involved is colored by this question of power.

The ostensible battle was whether or not some of the condors should be left in the wild. On the one hand, conservation groups such as Friends of the Earth and the National Audubon Society wanted as many condors as possible in the wild. For a long time, the U.S. Fish and Wildlife Service supported leaving a contingent of condors in the wild while building a captive flock. The main argument for this position was that having condors in the wild made it easier to protect habitat, and in the long run, good habitat was crucial for the survival of the species in the wild.

On the other hand, the Los Angeles Zoo, the San Diego Wild Animal Park, the California Fish and Game Commission, and several of the biologists on the project (though not all of them) favored capturing all the remaining condors and putting all the hope for the species, at least given the immediate crisis, in a captive-breeding program. Then, if and when captive breeding succeeded, young condors raised in the zoos could be released into the wild and trained to be free-flying.

In one camp were the ecologists, preservationists, and traditionalists. In the other camp was a new breed of biologists, ecologically rooted but manipulative and willing to intervene in natural processes.

Of all the people I met who favored keeping condors in the wild, Jesse Grantham was the one I got to know best. We spent time in the field together, and in my basic predisposition I shared many of his feelings about condors and the wilderness. His main arguments: The best way to save condors is to save habitat for them; you don't want to interfere with the traditions of the condor (he spoke of the condor's "culture"); and you don't know what could go wrong if you intervene with those traditions.

Immensely dedicated, Jesse organized his life around the unpredictable schedules of the condor, chasing after the birds all day and remaining on call for any emergency. One person on the project called him an "old-fashioned notebook naturalist." When a condor died not long before I met him, it was Jesse who went looking for it in the wilds and found it. And in an almost personal way, he was extremely worried about AC-3, trying to decide whether to risk going after her to see what was wrong, knowing full well that she might die from the stress of the handling.

Jesse was a quiet man whose heart had been torn apart by the animosities inside the condor project. It showed in his hazel eyes, pained and battle-weary. Though he spoke softly, with a voice that sounded like low blues from a sax, a defeated sorrow hung about his curly brown hair and his thick beard. No one who touched the condor project came away without wounds.

One time, when we were out in Jesse's pickup, hauling around stillborn calves from a local dairy (he put them out for the condors, so the birds could have "clean," or lead-free, carcasses to feed on), he made clear what was at stake for him with the California condor. It was much more than a single species. He was fighting for wilderness and for a way of life that included more wilderness.

We pulled up to a place with a view. In the winter, the condors hung out on the ranches in the foothills of the Sierra Madres, foraging for dead cattle, and we were skirting the Hudson Ranch, famous as the remaining condors' preferred winter foraging grounds. Below us lay the massive and heavily cultivated San Joaquin Valley, lost to cotton fields and oil rigs. Above us rose the tangled peaks of the Sierra Madres, cut by roads but stern and untamed still, scrambling out of the sea of agriculture that tried to choke them.

"What I envision," Jesse said slowly, "is that we manage all these lands as they once were. You know, we could release antelope and tule elk out here. They've been extinct in the region for a long time. And we could have a big preserve for other local endangered species like the San Joaquin kit fox and the blunt-nosed leopard lizard."

Throughout most of this century, the strategies that informed the work to save the condor, and most other species, have focused on what is considered to be the central problem endangered creatures face: loss of habitat. (Not everyone, however, agrees that habitat loss is the main or even an important cause in the condor's historical decline.) The condor's range once spread north to British Columbia and east to Florida. Fossils of condors have been found in the Grand Canyon. But its vast range had been severely reduced; at that time, all that was left was this small remnant of ranchlands north of Los Angeles.

For Jesse, the condor is more than a bird. It represents an ancient heritage of large spaces and unbroken stretches of time. The most numerous fossil remains of the condor have been found in the La Brea tar pits in Los Angeles, where the giant birds would get sucked in as they scavenged on mastadons and mammoths, dire wolves and saber-toothed tigers. Of course, this romanticizes the image of the condor, since robins, too, have been found in the tar pits. Still, to modern people, crowded by the millions in cities like Los Angeles, the condor came to symbolize both wilderness and prehistory. We picture condors soaring above the dry deserts of the west since the Pleistocene.

Here is the imaginative foundation of the fight to save the condor by leaving it in the wild, by preserving its habitat, by keeping its prehistoric traditions unbroken. It is really a romantic attempt to let the future catch up with the past.

The very idea of condors flying over wide, dry plains in an ancient landscape evokes both nostalgia and emptiness in me, a sense of all that seems missing from my own life. My days seem by comparison disjointed and monotonous. My minutes are small, broken apart, and repetitious; my spaces, all mapped and measured; my times of fullness and connection, few.

In the condor I see not so much another time, not just a prehistoric time—I see another *kind* of time. Jesse and I spent several hours, for example, watching for condors from the road along the Hudson Ranch. We were at a spot where, five years earlier, I had seen my first condor—a well-known spot, in fact, where birdwatchers frequently used to gather in the hopes of glimpsing a rare condor. From half a mile away, I had watched a condor appear on afternoon thermal winds, cruising above the brown grasses, dwarfing a raven flying nearby. Now, as Jesse and I looked out over the heaving and sagging hills, involuntary memories of that first condor kept clouding my mind, and the past kept inserting itself into the present moment, like palimpsests and overlays. Time is not always successive, moving forward from one point to another, with the past left behind as we enter a self-contained present, as if we were riding a train from one station to the next. My experience—watching for condors and picturing my first condors in the empty scene of the Hudson Ranch— reminded me of the kind of time in Marcel Proust's *Remembrance of Things Past*, where the past lives in the present; the two are not separate but alive and simultaneous. I see time not as a straight line through my life but as circles and spirals, full of repetitions and relived moments, the past always with me.

No matter what happens with the captive-breeding project, removing the last condors from the wild was a defeat for this vision of nature. No matter how successful the condors are in the zoos, even if they are someday released into the wild, capturing them was a failure for an America of wide spaces and slow time.

IV

Somewhere along the line, our dreams for the condor got tangled up in powerlines and windmills, lead bullets in deer and cyanide baits for coyotes, potshots from guns and collisions with airplanes —a gruesome and disparate list of the ways condors had been dying in the wild no matter how we tried to protect them. The mountains and ranchlands north of Los Angeles may have looked like the stuff of dreams, but not everyone was convinced that it was good habitat. The fact that it was the last of the condors' range does not mean that it was the best place for them. It may have been terrible habitat, full of dangers, but all they had left. While Jesse and I looked for condors from the road near Hudson's Ranch, picturing it, transformed, through the alchemy of imagination, we did not see any. AC-3 and her mate were there, but they stayed down in the gorge, not visible from our spot on the road. Their stillness was an enigma.

Everyone, including Jesse, agreed that the long-term hope for condors did not rest in the wild population. Some new logic would be required to save them. Noel Snyder, more than anyone, articulated and campaigned for a new approach to saving the condors. I heard him described (by a director of the Condor Research Center) as the single biggest problem for the condors. I also heard him described (by many of the biologists on the project, who looked to him for leadership) as the number-one reason that the California condor still has any chance for survival.

Noel's frayed brown sofa, in his living room in Ojai, California, was famous: Condor biologists called it the "Coma Couch." For eight months, Noel had lived on that couch—eating there, sleeping there, working there. The preceding March, when the six condors had turned up missing, presumed dead, Noel had begun arguing vociferously to capture all the remaining condors, making matters so tense that the Fish and Wildlife Service tried to transfer him off the project. The stress of the condor program, however, brought him a heart problem, so he took a sick leave and retired to his living room. From there he kept in daily contact with the biologists in the field, many of whom stayed fiercely loyal to him, feeding him information on the birds and taking direction from him. He had become a biologist-in-exile, and the Coma Couch a covert mission control.

I met him the week the U.S. Fish and Wildlife announced it was issuing permits for the capture of the remaining condors. I lived at his house. Noel had fought for ten months for this change of direction, so the announcement was a personal vindication for him. He was convinced the captive-breeding program would be the best thing for the condor.

For several days, he only rarely left the couch. Wrapped in an old sleeping bag, gray hair and beard disheveled, forgetting to change his red socks or brush his teeth, he would wake up, say, at 2 a.m., work on a manuscript on his blue IBM Selectric on the coffee table, and then fall back asleep. Or he would talk on the little blue phone to biologists. That phone, with about thirty numbers taped to it, was his lifeline. In eight months, he had run up $2400 in long-distance calls.

A concert cello student when he was in college, Noel was now slightly overweight from enforced sedentary living. But his sharp facial features, quiet voice, and incisive mind contributed to his powerful charisma. Amid unrelenting talk about condors, I slowly came under his spell. Here was the condor counterculture, and Noel its guru, its beat intellectual.

When he came to the Condor Research Center in 1980, he already trailed an intimidating aura as a pioneering biologist. He was well known for having developed the methods that saved the Puerto Rican parrot, as well as for his major research on the decline of the snail kite in The Everglades. Now, years after the condor crisis, he

is working in Arizona, reintroducing long-vanished thick-billed parrots into the Chiricahua Mountains.

The plan Noel developed for the condors was innovative, unor-
thodox, and obviously controversial—it cost him his job with the
U.S. Fish and Wildlife Service. Giving up on the wild population,
he proposed a three-part plan: First, catch all the condors and put
them into the two zoos. Second, breed them in captivity (which at
that time had never been successfully done with condors). Third—
and this was by far the most unusual and interesting aspect of the
plan—release the condors produced in captivity at some indeter-
minate later date, and train them in the wild so that they would not
fly out of the remote mountains in the Transverse Range north of
Los Angeles.

In Noel's view, the condors' current foraging habitat in the
ranchlands around the San Joaquin Valley was the principal problem. He took the plight of AC-3, a bird he knew intimately from
years of watching her in the field, as further confirmation that the
condors had to be removed from that area and prevented from returning to it in the future. It was killing them, and there was nothing
anyone could do to stop the birds from dying as long as they remained
where they were so visible, so accessible, so vulnerable. Noel called
it "terrible habitat."

In his throaty voice, with its steady, insinuating power, he made
it clear that a new "science" was at stake. "The condor has come to
be a symbol," he said. "It came to glorify wilderness, even though
it's out there feeding on ranchlands. It's given them a mystical status.
But boy, there's a real problem trying to manage a symbol. It's like
trying to manage smoke rings.

"We had to prove the condor was a bird."

From the beginning of the recovery project, Noel advocated a
more active approach to the condors. This engendered controversy
from the start: While project members were trying to weigh young
condors on their nests in 1980, one chick died from stress. The whole
debacle was photographed, achieved national notoriety, had preservationists screaming "hands off," and left Noel and others on the
project living under a poisoned cloud that they called "The Curse
of the Condor."

But Noel persisted. His biggest contribution for the condors came in taking eggs from the wild, moving them to the zoos, and hatching them in captivity.

For Noel, the main breakthroughs for the condor were based on what he called "good old-fashioned watching," which might seem a paradox for a "manipulator," except that it shows his grounding in strict observational techniques.

To help an endangered species, you can try either cutting down on deaths or improving birthrates. Like many of the most endangered species, condors are slow breeders. It takes six or seven years for the birds to reach sexual maturity; then they lay one egg per nest, and are thought to breed only every other year. This reproductive strategy, which biologists call "K-selection," depends upon a high rate of survival for the young. Other animals use an "R-selection" strategy, producing many young who have a high mortality rate.

But condors were losing about 50 percent of their young. They nested in caves on mountainous cliffs, taking turns guarding the nest. While one parent remained with the chick, the other parent foraged, bringing the nestling food, which was vomited up. The arrangement was not without hazards. One pair, for example, were notorious as careless parents, often jostling as they switched turns at the nest. In 1982, in a transfer at the nest, the parents knocked their egg off the cliff, and it became food for ravens.

The conventional wisdom said that condors never lay a replacement egg, but forty days later, these two condors did lay a second egg. This not only proved that condors "double-clutch," as it's called, but also enabled Noel to get permission to pull condor eggs, take them to the zoo to hatch, and stimulate the nesting condors to lay additional eggs.

In 1984, Noel led a team into the mountains to take eggs from condor nests. The most dramatic lift came from the nest of AC-2 and AC-3 on Madulce Peak in the Santa Barbara Mountains. The pair had nested in the middle of a huge cliff, a "death ravine" as one person put it jokingly. Noel led the crew "into the abyss, an epic from start to finish." It took two ropes to get down to the nest, and then, once they got the egg, their helicopter almost crashed.

Eventually, Noel demonstrated triple-clutching. The eggs were taken to the zoos where, after hatching, the chicks were raised using specially designed, hand-held puppet parents. AC-2 and AC-3 provided six eggs to the captive hatching.

In 1981, there had been only one captive condor, the famous Topa Topa. Between 1983 and 1986, sixteen eggs were removed from the wild and thirteen baby condors hatched successfully. In addition, four nestlings were removed from the wild between 1982 and 1984.

When the first chick hatched in a zoo, the mood among condor biologists lifted and public opinion changed. For the first time in a century, there seemed to be real hope that something might be done to save the condor.

In building the captive flock, Noel and the zoo biologists were looking ahead to captive breeding as well, so eggs were brought in with an eye to enhancing the genetic diversity of the population. But when the six condors vanished in 1985, as far as Noel was concerned, a new stage in condor recovery had arrived. The captive flock could no longer be considered a second strategy for the condor's future. It *was* the condor's future. All the birds had to be brought in to give the captive-breeding program the greatest chance of success, as well as to protect the wild birds from the vagaries and dangers of life in the wild.

Captive breeding evoked real worries, not only among preservationists, who saw it as abandoning the dreams for wild condors, but also among biologists like Noel. It has obviously had precedent, being used with other species like whooping cranes and Puerto Rican parrots. Yet in most cases, endangered species had been very difficult to breed in captivity, though in each instance, a closely related species, the "surrogate species" as it's called, would do so fairly easily. Sandhill cranes did. Hispaniolan parrots did. But for some unknown reason, perhaps genetic, the endangered species did not. Andean condors had been bred in captivity at that time, but not California condors. Though Noel was confident, there was nevertheless a big risk in putting all the hope for the most publicized endangered species in North America in a program that had had no successes yet.

When the baby condor hatched on April 29, 1988, in the San Diego Wild Animal Park, it was the first condor produced by captive breeding. Named Molloko, it was additional vindication for biologists like Noel, as well as 6¾ ounces of hope for the condor species. Mandan, the program's second chick, hatched in April 1989, bringing the population of condors to twenty-nine.

The really controversial, and totally innovative, part of this program to save the condors entirely through captive breeding will not come until the late 1990s, once a flock of young condors has actually been produced in the zoos: releasing condors into the wild. The man in charge is a young biologist named Mike Wallace. His plan is stunning, even a little scary.

Noel encouraged me to meet with Mike, then fresh from years of experiments in Peru, where he had been releasing Andean condors into the wild. On the basis of his work there, as well as what he knew of experiments with griffon vultures in Europe, he was utterly convinced that he could not only release California condors into the wild successfully but also train them to do exactly what he wanted.

His plan is to establish a new condor habitat and not to let the condors go back into the grasslands, oak savannas, and ranches bordering the San Joaquin Valley, where so many had died from shootings and lead poisoning and other causes. He and Noel chose a remote site, Hopper Canyon in the mountains of the Sespe Condor Sanctuary, already set aside and protected for condors. After he releases the captive-bred condors in Hopper Canyon, Mike will use what he laughingly calls "carcass management" to teach them to fly, eat, and breed completely in that area. Using birds that have never before been wild, he'll teach them a new tradition in the safety of a remote wilderness.

Mike's experiments with Andean condors in Peru were impressive. A smart young Ph.D., he had spent three years of doctoral research dragging carcasses up and down beaches and mountainsides, testing whether he could control the condors' movements. Not only did he teach them to fly, but he claims he could lead them anywhere he wanted with his carcasses. Since everything the condors needed was provided, they had no impulse to range abroad. Mike's confidence was unbounded. He had, he said, "complete control over them."

Opponents like Jesse Grantham are left someplace between a

guffaw and a howl at the prospect of trained condors. Condors are birds of flight, they say. That's what the 10-foot wingspan is all about—the birds think nothing of flying 100 miles at the drop of a hat. Not only would trained condors in Hopper be a desecration of the birds' dignity—one step away from feeding them Kibbles and Bits on the Johnny Carson show—but it's close to biological hubris to believe we can totally control such a bird. And if the grand experiment fails? If the California condors fly out into the ranchlands again—habitat that will have grown more developed in the intervening years, more dangerous because we will have abandoned the effort to save it? Then what?

For Noel and for Mike the challenge, and the prospects, are inspiring. As Noel said, "I think we've crossed a threshold. Scientists versus environmentalists? Maybe. This is an exciting new groundbreaking experiment. I want the birds in the wild as much as anyone else, and some of those future wild populations will be less managed and wilder than what the species has enjoyed for many decades."

The California condor is a major precedent for the manipulation of endangered species as the best hope for their survival. Power has changed hands from preservationists to interventionists.

In the end I became convinced, reluctantly, that Noel's blue phone and Coma Couch may have saved the California condor. I came out of the research deeply disturbed yet extremely excited.

Disturbed. Make no mistake about it. However sincere the motives of Noel Snyder and Mike Wallace on behalf of condors (and they are as pure as anybody's, I think), a major social statement has been made in having to put all our condor eggs in the zoos' basket. Not only are we acquiescing to a notion that creatures are objects to be manipulated, things we control with our heads rather than respond to with our hearts, but we are also doing something else. By giving an increasing role to zoos in rehabilitating endangered species, we seem to be saying that the place for animals in our culture is in confinement. We may call it safety, but who is it we customarily place in confinement? Outcasts. Misfits. The marginal. The mad. The criminal. This is merely one more sign of the increasing alienation of modern society from the parts of nature that reason cannot control. Animals no longer have value to us as symbols from the beyond; they are allowed merely to live on handouts along the borders of our lives. They are creatures for show only, things to look at but not to know in any personal way: "monsters" on display—and the etymology of monster is "to show." Condors in captivity, condors on a carcass leash: one more loss in the fight for the values of wildness and wilderness.

Yet, excited. Imagine some day, maybe twenty years from now, hiking in the dry mountains and chaparral of Los Padres National Forest. It is late afternoon. Hot air rises from warm rocks. A California condor, wings fixed and riding an updraft of air, floats beneath the eternal face of the blue sky. And maybe some of AC-3's genes are in that condor; maybe one of the eggs Noel lifted from a nest ultimately produced it. My heart would soar with that bird, filled with joy at what humans can do for condors.

V

But wait. Can you smell the stink of some corpse rotting in a corner? Some secret on the edges of the shadows, just out of the bright lights of science? I want to let my eyes get used to the darkness on the edges a bit, shift my ground some, try a slightly different perspective on this whole tempest.

What we don't see, by letting the biologists set the terms of the debate, is that some larger social collusion has taken place. Though I get impatient with biologists and their approach to animals, my complaint is not really with biologists. People like Jesse Grantham and Noel Snyder have made heroic efforts on behalf of the condor, as other biologists do for other endangered species.

When the last of the condors was captured, society suffered a loss—the loss of the wild condor. But something else achieved a victory. The winner was biology. It confirmed its right to define for us what a bird is.

We should ask ourselves, why we are willing to allow biologists to tell us what a condor is, plan its future, control its fate. It is hard not to suspect that there is something convenient for us in this arrangement, something useful to us in having the condors in the zoos, something all right in the failure of the condors in the wild. I find a metaphor in the capturing of the condors. We gave the condor to the biologists to study and to save. They extorted some truths from the condor: what it needs to live, how it lays eggs, what it might do when released. But the price of that knowledge has been the extinction from the wild of the old and ancient bird.

This is taking us closer to the secret of the condor. Its extinction is not a local problem, the result of shooting or lead poisoning or even habitat loss. Nor is the epidemic of extinction in our time a local problem. It is a necessary consequence of our way of seeing nature and relating to it. The fact that we define extinction largely in biological terms, instead of, say, social or psychological terms, is simply an expression of the way we see nature—an expression of what caused the problem in the first place. Biologists may have some stunning successes with endangered species, and they may even save the condor. I hope so. They have enabled the brown pelican to make an exhilarating comeback and have helped the peregrine nest in tall buildings of our cities. But these are isolated achievements in a landscape of much more sweeping loss and extinction—thousands of plants and animals in danger. During the twentieth century, mammals and birds have disappeared at the average rate of about one species per year.

Biologists may be able to document the problem, but surely we are not so naive as to think they can solve it. They cannot solve it because the scientific way of understanding nature has helped cause the problem.

The explicit debate about the condor's future—whether or not to capture the bird—masks a darker truth about the fight. Underneath, it was about science as power.

The late French philosopher Michel Foucault exposes, attacks, and studies this substrate of power under knowledge everywhere in his work. In *The Order of Things*, a brilliant and iconoclastic analysis of Western epistemology, he is not primarily concerned with a traditional history of science. Rather, he calls his work an "archaeology of the human sciences." He looks not so much for the processes and products of the scientific consciousness, as most historians do, as at what has eluded the scientific consciousness. This is the unconscious of science, implicit in the way science chooses to talk about nature.

The rational and the empirical are what Foucault calls the "governing codes" of our culture, expressed through our science and used to give our particular form of order to the world we inhabit to make sense of it. But it is possible to look behind those codes to truths that cannot be said by the governing code, to an "unspoken order." In *The History of Sexuality*, Foucault finds beneath our rationality the "omnipresence of power." Power, he says, "is not an institution, and not a structure; neither is it a certain strength we are endowed with; it is the name that one attributes to a complex strategical situation in a particular society."

The knowledge generated by science, for Foucault, is one expression of our urge to achieve power over nature. Foucault writes that "relations of power" are not exterior to knowledge, the results after knowledge is generated when we try to apply that knowledge. Rather, power is "immanent" in knowledge. Every kind of knowledge is the expression of a kind of power. Every kind of knowledge creates divisions and inequalities and dark recesses—a certain landscape of its own, a certain politics in the relationship between what is known and what is not.

Power is rational, its tactics often quite explicit. But power also always wears a mask so that part of it is hidden, even from the wearers. We camouflage ourselves in our explicit, good intentions and deny the work of power. We want to save the condors, we say, by putting them in zoos, but we don't admit that this actually gives us complete power over the condors. In this way, to use Foucault's language, science becomes the ruse of reason.

How does Foucault's analysis of power apply to the condors, to endangered species, and to the politics of conservation?

Following Foucault, we should be able to ask what deeper objectives are served by the kinds of knowledge we generate about nature. Are power relations at work in the posture we take toward condors? What kinds of domination are served by our knowledge?

Ecologists, for example, were able to exploit their knowledge of condors, to transform it into a symbol, as part of their resistance to the dominant culture that almost destroyed the species. Knowledge about habits and habitat gave a scientific basis for a rebellion against the exploitation of nature. And the more recent manipulative biology is part of a strategy as well. It exists not as pure knowledge but as part of a means to gain control over the bird—ironically, to save it. But the very saving of the condor means restricting it to zoos or to carcasses we place and move about. This is not happening only with condors. We now save many endangered species by losing them.

What has happened in the particular case of the condor is a symbol of what is happening more broadly with other endangered species. By trying to control animals through "bio-power," we create failures while we create successes. The focus on single species, the emphasis on identifying problems, on generating solutions, and on gaining control over nature—these are the forms of power expressed in our current approach. Yet this very desire to exert power over nature has informed our scientific approach to nature since the seventeenth century. The methods we have used to save endangered species must fail, in the larger sense of all endangered species, because the scientific approach to animals is part of the cultural mentality that created endangered species.

The problem of endangered species is only superficially a biological problem. Endangered species are the inevitable expression of our power over nature. We have invented animals as biological creatures, and at the same time we have turned these creatures into strangers. Putting creatures in zoos to save them is a strange formula. It is a way for us to confirm our alienation from the beast—quite literally we ensure that it is no longer *inside* us but living in a separate, designated space, a beast we have isolated. In this way, the animal in us finds not its expression but its cure. We hand it, like the condor, over to the zoos, put certain people in charge of it, and lose touch with it. In the zoo, we make animals strange, rare, distant; we even confer upon them a special status. The thirty-two remaining condors (as of 1989) now live guarded in the zoos, and we cannot see them. We lose touch with a creature's secrets. But at the same time, and this is the central paradox, we also demystify the creature. The condor lives the life of a secret, but the zoo, run by vets and bureaucrats, steals its strangeness and makes it a creature we can train.

The condor shows the double face of our science: Where there is power, there is also impotence. And where there is light, there is also shadow. The whole dilemma is summarized for me in the Greek myth of Prometheus. A culture hero, Prometheus conferred upon a pathetic humanity the gift of fire: arts, crafts, and knowledge. It was an act of power, an act of rebellion against the gods. For that act of presumption, for the light of Greek science, Prometheus was punished: He was chained to a rock in the mountains of Caucasus, and his liver was ripped out and eaten every 1000 years by a vulture.

As does Prometheus, the vulture teaches about the limits of our power. And so I'm left thinking we need to turn to other places for our answers to endangered species. We can't expect biologists and bureaucrats to solve our problems. We need a revolution of consciousness. That's why I believe in desire and passion, make my attacks upon the sovereignty of reason, look for answers in the shadows and the gaps of our knowledge—and in endangered species. In our failure with the California condor, I feel shifting in a ground that usually seems so solid, so real, and so accustomed that it cannot be questioned. Maybe the ground is stirring beneath us. The last range of the condors, after all, was right over the San Andreas Fault.

VI

The closest I have ever come to heaven was the time I flew with condors. I turn to immediate experience to break through the overlays of power and politics, intellect and interpretation. My experiences dissolve over time in my consciousness, changing shape and meaning, suffusing through my mind and changing the way I see things. In one day of flying, I came away with two experiences that reminded me what we'd done to condors and showed me what we'd lost with them, experiences that left me soaring on the winds at 10,000 feet.

The pilot was Buck Woods, a young biologist who specialized

in radiotelemetry work from planes. The incandescent southern California morning burned cool, lovely now in the long shadows and hints of heat, but I could already feel the wallop of a hot afternoon wrapped in the chill air, like a fist inside a glove. Under a loose T-shirt, Buck's brown skin showed the muscle tones of a weight lifter. A man of his own opinions, he had moved to New Cuyama, where I met him, to get away from the politics of condor management. He took care of his Citabria ("airbatic" spelled backward) on a beat-up, abandoned old runway about 20 miles from the Hudson Ranch. The plane was small, blue and white, and had room for two, front and back, like an old warplane. It maneuvered deftly with the condors in the air.

We had two missions for the day: In the morning, help the crew at Hudson Ranch with AC-3; in the afternoon, follow two other condors, AC-8 and AC-9, to their roosts in a wild section of the mountains called Agua Blanca.

I squeezed indecorously into the plane's backseat, and Buck skimmed us over the San Andreas Fault in the Carizzo Plains, on our way to the Hudson Ranch. The fault lay like a huge gash in the dry dirt. We flew across the plains, above the bumping foothills, over cattle and fences, and circled over a bucket of green against a canvas of winter brown—Bittercreek Gorge.

Buck talked over the radio to the biologists on the ground, who were standing beside their pickups. They hadn't been able to locate AC-3 today. "They want me to make a low pass, see if the birds will get up," he said to me. "They're gonna hike down into the gorge."

We made a slow, wide arc out over the ranch, gained speed, and came in on the gorge low, engine cranking louder, cattle gaining speed below us, pressure mounting on my chest and keeping me from breathing. Airspeed, 85 knots. Altitude, 30 feet.

Then, in a silent swoosh, the ground crashed away below us into a tumult of stone and a jungle of green. We bellied into the gorge, our bottom sagging into the cool and heavy air. Buck pushed the Citabria's nose downward, aquaplaning over the sea-green trees, skittering over sharp pines and rounded oaks. It was a rich and delirious plunge.

I spotted the rotting cow.

And then, against the shadowed bottom of the ravine, I picked up another shadow, except this one was flying. It bolted down the gorge, staying low and keeping to the shadows. Big. The lavish morning sun flooded even the shadows of the gorge with a reflected shine, a luminous dazzle.

"It's AC-3," Buck said. He could tell from his radiotelemetry. Each condor had its own frequency.

That was my last look at AC-3. I didn't see much in the brief pass: I have more of a dream-image in my mind than a sharp memory. I remember her broad-backed and wide-winged, flapping, a gray mottle on her wings, feathers on their tips, like fingers, turned up in the drafts of wind. And her black body, disappearing into the shadows, vibrant with a glassy, melanistic sheen.

"That musta scared hell outa her," Buck said. "We hate to do that, but wanna see if she'll fly without havin' ta catch 'er."

Buck wanted to believe that the air was too heavy for soaring, so she chose to sit on the ranch. Or that she had been feeding on the cow carcass. "AC-2 and AC-3, they're this stodgy ol' couple," he said. "All they did was fly back and forth, Madulce Peak to Hudson Ranch, nothin' very exciting. But they're the big egg producers. And when they're flyin', they've been together so long, it's like they can read each other's mind. Bank in unison, do everything together."

But bad air for AC-3 was wishful thinking.

The biologists decided just three days later to try to catch AC-3. They had to learn what was wrong. But despite their high anxiety, they didn't capture her until January 3, 1986. For almost a month. she had not left Bittercreek Gorge. She was so weak that when Jesse and the other biologists finally captured her, they just walked up and grabbed her.

Their worst fears were confirmed by tests. The high levels of lead in her system, present for at least two months, had slowly debilitated her. The lead had poisoned her system. She could eat but not digest. (Lead paralyzes the peristalsis, and the gut freezes up.) AC-3 was starved, scrawny, dying. Her crop was in fact engorged. There had been so much damage to her nerves that she was not able to swallow any of the meat in her crop. Some of the carrion she had eaten had been in her mouth for two weeks. It just sat in her crop and rotted. So in addition to poisoning and starvation, she was choking on the guts in her maw, green and gross.

AC-3 was taken to the San Diego Wild Animal Park, and veterinarians treated her for lead poisoning. They laid her out on a white antiseptic table, stuck plastic tubes in her to feed her intravenously, and pumped her system repeatedly, trying desperately and futilely to clean her out. Fifteen days after her capture, on January 18, 1986, AC-3 died.

AC-3 was the last wild condor to die. She did not precisely die *in* the wild, but she might as well have. She was the last successfully breeding female in the wild, and her eggs and babies provided much of the core for the captive flock of condors. She was killed by lead poisoning, probably picked up from bullets in the carcass of a deer which a hunter had shot but lost.

AC-3 was a very important condor in life and death. She revealed enormous amounts to biologists about the natural history of condors, and through her eggs the condor still has a chance for survival. But AC-3 was also the epitome of the horrors suffered by the California condor for a century at the hands of humans.

We were impotent in our attempt to save her—even with crews of biologists watching her, even with a plane tracking her, even with vets treating her.

At least, though, I can memorialize her now, here. It's a small gesture, but I remember her, barely able to fly, preening in a tree.

None of this very particularized sorrow had yet taken shape, however, when we made the pass over AC-3 and I saw her for the last time. Her fate was still uncertain, folded in the future, hidden in our worries.

Instead, I had the sense that we had just had a hint of what was possible when flying with condors, a brush with gut-rolling exhilaration. It turned out to be a foretaste of the afternoon.

After landing at Edwards Air Force Base for a burger, we flew into the Tehachapi Mountains on the east side of Interstate 5. Up on the rolling, potrero-marked crest on the Tejon Ranch, amid parched grasses and the inkblot stains of dark oaks and junipers, several biologists were trying to catch a condor.

The strategy, adopted from California Indians, was to hide in a pit next to some bait, usually a cow carcass. When a condor hopped up onto the carcass, the biologist would reach up out of the pit and grab the bird by the legs. It's a slow way to catch condors. It would take biologists sitting in pit traps almost a year and a half to catch the last six condors in the wild. That's a lot of waiting, but it is also much safer for the birds than using a cannon net.

When Buck and I got to the trap site, two condors were flirting with the biologists. One condor was circling in a rimrock canyon, just off the crest, and another was on the ground, tantalizingly close to the carcass by the pit. The bird in the air was AC-8, a female. She faded toward the ground while we watched her, coming to a running stop near some oak trees, very much like a hang glider landing.

AC-8 had been a reliable breeder, but in the rash of death during the previous winter, she had lost her mate. Now another condor had been following her around, hanging out with her at the roost. AC-9 was a young male who'd never mated. Buck loved him. He loved to watch AC-9 make his move on AC-8, and he hoped they might become a breeding pair. Buck was tracking these two condors to see if they might be pairing up, and we were going to follow them to the roost that night. "Put 'em to bed," as Buck said.

AC-9, the condor by the trap, was barely an arm's reach from the carcass. With a hulking dignity, young and daring, he strutted toward the cow; he was magnificently clumsy on the ground. He leaned forward, hooked a chunk of meat with his beak, and jerked at it. For a while, I thought we were going to see a condor get captured. Instead, AC-9 just kept us on a tether, teasing us by getting close but never close enough, jerking us around as much as the cow. AC-8 played coy, looming spectacularly in the background shadows of the live oaks.

By late afternoon, the two condors decided to take off. They were so big that they needed a cliff, good winds, and a running start.

With their wings flapping for balance, the ungainly pair lumbered down an open slope, making toward the cliff, and then lifted themselves into the air currents of the mountains. The warm air of afternoon carried the condors into an easy, dreamy glide, so different from their cumbersome movements on the ground.

They headed west, toward Interstate 5, AC-8 leading the way. We trailed discreetly, and I realized I had now seen five of the last six condors.

Buck yelled into the intercom: "I've followed 8 around for hours. Once in a while, I'll catch her in a cruise and I'll just go with her 'til I run out of gas. I've flown wing-to-wing with her at 60 mph. Never flap. And 9, chasin' after ol' 8, whatta stud."

South of Bakersfield, the freeway starts a long climb out of the low heat of the agricultural valley, 20 or so miles of unremitting grade. Trucks in the right lane slow to a crawl, and when I drive the "Grapevine," as it's called, I find myself leaning forward into the wheel, trying to give my car a little added momentum.

The condors took us out over Interstate 5 at the Grapevine. The freeway slashes a deep chunk out of the mountains as it rises to the summit of Tejon Pass. On either side of the concrete river below, cliffs of ragged, torn rocks rise hundreds of feet. Coming in off the ocean, winds whip through the artificial canyon of the freeway, and the hot sun heats the concrete and cliffs through the course of the day. By afternoon, the air in the canyon of the freeway is alive with radiant heat, swelling thermals, slicing winds. When we took the Citabria out over the freeway, the air had an almost physical jolt in it, a powerful surge up.

The air over the freeway must have been the stuff of dreams for condors. They did not cross the freeway. Instead, they began circling, wide and slow, wings held steady, tilting into the turns. One of the condors rolled left, the other rolled right, spiraling upward in a double helix, as if invisibly connected to some power in the billowing air.

Buck leaned the Citabria into a tight spiral, too, and we rose with the condors. Slowly, almost deliberately, like partners in a dance that is unrehearsed but that unfolds with a kind of inevitability that looks calculated, all of us flew higher, leaving the freeway behind, forgetting the cars below pouring by the thousands in and out of Los Angeles.

"They're flying real tight," Buck said. I could now see the advantage of the Citabria, insinuating itself into the condors' flight with such intimate aerial aerobatics.

We rose above the highest of the mountains. The San Joaquin Valley shimmered like a mirage under a heat haze to the north. Los Angeles to the south blinked through the smog. Still, we kept circling higher with the condors, the views getting more expansive, the spirals growing wider in a dilating euphoria.

The California Indians revered the condor, conferring on it the power to transport humans into a different dimension. Many tribes celebrated a festival called *panes*, in which a condor was sacrificed to a god and its skin, removed in one piece, was buried with seeds. The condor was thought to be transcendent, able to come back to life. In fact, the Indians believed the same condor was sacrificed every year. In a tale from the North Fork Mono people, the condor was so powerful that he would carry people who slept in the open up into the skies. Then he would set them free in "skyland."

I don't usually like seeing wildlife from a plane very much: It's often the easiest way to see rare creatures, but it's not an intimate way to see them. With the condors, however, the experience in the plane was building into something remarkably different. I was with them in their element. In the air, they weren't cumbersome. They were at home. And they were lifting me, as in the Indian tale, into a higher space, a paradise of flight and wide-open skies.

At several thousand feet, the condors banked out of their climb and veered south, following the line of the freeway. They held their wings firm, flapping rarely, in what Buck called a "flex-glide." In the plane, we stayed about 100 feet off the wingtips of AC-9. His wingspan looked almost as broad as the Citabria's, though it was actually almost a third the size of the plane's 35-foot span.

Below, the cars creeping along the road seemed increasingly meaningless. I thought of all the times I had driven that stretch of freeway, craning to look up through my windshield, hoping to sight a condor. For the freeway cut right through the heart of the condor's last range. That was the culture below, and it seemed from up here suddenly earthbound, blind, oblivious to the drama unfolding in the skies above.

In the air, the condors had another culture, built on winds and air and wings. Next to us, AC-9 cruised on giant wings, body immobile, steady and strong. The feathers at the tips of his wings blew and fluttered as he glided through the air. His head swiveled, the way a modern dancer isolates and moves a single body part. I looked out of the plane window, watching his red head rotate while he flew, looking below, looking sideways.

And looking straight at me.

There's nothing like being transfixed in the gaze of a spectacular wild animal. Always it's a shock to me—the sudden recognition of strangeness. In the condor's stare, I felt a disorienting self-consciousness that came with losing my role for a precious moment: I was no longer sure whether I was the seer or the seen. I got a fleeting sense of what I must look like to the condor, both of us made visible in the same light of day. The gaze also forced me to enter the picture, to occupy a place in the skies with the condor. In the look of the condor, I recognized a part of me that had existed unknown to myself.

I could also see in the look that the condor has a world and a life of its own, that it will always be strange to us. This was the condor of the skies, a magnificent soarer, looking down upon the monotonous and endless traffic of our lives.

I couldn't help seeing in the flight of these two condors some destiny in the making, and it was deeply moving, because a destiny is always a mystery as it unfolds. Would they breed together? When would they be captured? Would their offspring someday be released in Hopper Canyon? Will the condor survive?

If you could have seen me in the plane, you would have laughed. I was bouncing up and down in the backseat, screaming for joy and pounding Buck on the shoulders in ecstasy. Buck was laughing, happy to share his experience of condors with someone else who felt the exhilaration they afford. "Only four people in the world," he said to me, "have flown like this with condors. Seen condors this way."

After flying several miles up the Grapevine, the two condors turned west into the Sierra Madres. They cruised low along the hills and ravines, using the currents. We droned higher in the plane, and the birds dwindled into small shapes.

In the low December sun, shadows seeped into the gorges and up the sides of the cliffs and chasms. The day faded in a subtropical swoon of rich light—velvety reds, dusky peaches, heady yellows, a tinge of purple that almost hurt to behold. In the liquid evening haze, you could almost feel the air, taste it, drink it, but never get enough of it.

The two condors made their way above the wilderness, heading toward their roosting site at Agua Blanca. They would slide low along a ridge, feeling their way through the mountains on currents, invisibly connected to the contours of the hills. On the sway of the winds, they had an unbelievable grace, indolent and unhurried. At Agua Blanca, they again began to circle, staying just above the shadows, reeling in the fiery light. Now and then, they would slip out of the light, down a ridge, lost in a black swatch, only to catch another thermal and burst back into the light, like buoys breaking the surface of dark waters.

AC-8 and AC-9 spun together in lovely pirouettes above their roost. The sun burned red along the horizon, and the condors hovered just above the canyon. The late rays of light splashed off their broad wings. Emblems of the species, polished and brilliant and unforgettable, the two condors hung just above the dark abysses below.

FOUR

A PANTHER IN A SWAMP

I

It is comforting, I suppose, to realize that we have invented animals in the last few centuries. In the case of the condors, our way of seeing condors as animals, of constituting them as creatures, is part of the problem the birds face. But it must be possible to create other ways of seeing animals as well. New forms of consciousness, new ways of coming upon the animal in the wild, must lie as buried potentials within us.

How do I begin to re-create my world in a way that goes beyond merely imposing some new imperative upon myself, some new dictate, some new morality—even if it's garbed in the clothes of an environmental ethic? I want to see animals and my world in a fuller way. I want to create a relationship with the world that comes from inside as well as outside, that feels real. And I want to do this not just with my head, either, but with my heart also, drawing not only on my trained and civilized feelings but on the darknesses too.

I want to know something about life—my own and that of the other creatures I encounter—through my experiences, not filtered through the images and ideas of someone else. But it is hard to know my own experience. Francis Bacon warned, in the seventeenth century, of four kinds of error the human mind is prone to. He called them "idols," and two of them—idols of the den and idols of the theater—make us believe that things are as we have learned them from words and images and theories. The tyranny of images keeps me from believing my own perceptions, seeing my own experience, having a place in my own life.

This was certainly the case when I chased after a Florida panther in a swamp.

When I think of panthers and mountain lions, I think immediately of the great romantic painting done by George Stubbs in 1770, and now in the Yale University Art Gallery, *Lion Attacking a Horse*. Nature as the place we turn to for the grand, the sublime, the dramatic—revealed in moments of adventure and terror and violence.

I think, too, of the other romantic vision of nature, one that is still a staple of nature writers: the lyrical. Nature as the home of the alienated heart, a place where we can be reintegrated into some larger whole. William Faulkner expresses some of this lyrical sense of nature when he writes of the southern swamps and the wildlife of Mississippi. In "The Bear," he mentions panthers and wolves, as well as the great bear. In the chase to find the bear, a young boy faces toughness, triumph, and transformation. The swamp offers an initiation rite: Truth may be various, confusing, and swampy, but we can find it because, Faulkner writes, the "heart already knows."

What I experienced in the swamp was nearly a parody of a romantic nature, except that it was also sad. The panther I found lives someplace between an adventure and a travesty.

I joined a crew of biologists in south Florida who were doing research on the panther. Its scientific name is *Felis concolor coryi*. It is an extremely rare mammal. Mountain lions or panthers or cougars, as they are variously called in different parts of the country, are always secretive, perhaps the most difficult of American wild animals to see. Every photograph you've seen of mountain lions, on postcards or in magazines, was probably of a tame lion that was taken to an appropriate setting to be photographed—much of our nature photography is an illusion in more than one way.

At that time there were as few as twenty Florida panthers left. The official estimate, while I was in Florida in the winter of 1986, was twenty-seven. These panthers are hanging on in a small pocket of swamps in south Florida—in the Big Cypress, The Everglades, and the Fakahatchee Strand—and they are the only mountain lions still surviving east of the Mississippi. Once, the species was common throughout the United States, but it has been hunted down and exterminated as a predator much as the wolf was.

Extinctions in the United States and other temperate zones are still largely of populations rather than entire species. The Florida panther is a remnant population, the others in the east having vanished. It is a deeply endangered animal, perhaps unsavable, and is almost never seen except, ironically, in pictures as the official Florida state mammal.

II

Like me, the biologist in charge of south Florida research on the panther for the state, Dave Maehr, had never seen a panther. It was his first day on the job, but he had plenty of experience from his work with black bears. He was young, eager, blond. We were both in a way innocent.

Nine of us were on the expedition. The plan: to track and tree a panther in the swamp using hounds.

Sound romantic? How should I plot this story? What kind of narrative will shape the way I see this panther? I think I need a woman, one to fall in love with during the ordeal.

There was a woman on the expedition, the veterinarian Dr. Melody Roelke, of the Florida Game and Freshwater Fish Commission. Once we got to the panther, she would take charge, a female telling eight men what to do. For a story, the situation held great potential.

A couple of the guys on the team were willing to play good parts—young southern country guys, big, playing dumb but smart underneath, and full of jokes and laughter. Walter McCown and

Jayde. Jayde was angular and strong—and he refused to tell me his last name. Walter was tall and round—the night before, in the bunkhouse, he wore huge, fluffy, bear-claw slippers. Today, he wore a machete on his belt and a wristwatch with an alarm. The alarm went off as we were piling equipment into the swamp buggies, chirping out the melody of "Love Me Tender."

Both of them teased me as we got ready to go into the swamp, trying to scare me a little.

"Good day for cottonmouths," Walter chuckled. "I've seen twenty or thirty of those snakes in a half hour. So big, all they do is fall on ya and they'll keell ya." Both he and Jayde were laughing.

It *was* a good day for snakes. After a spell of cold weather, the sun was out: a sharp light, bright but carrying in the morning coolness a trace of vaporous softness. There was an implied heat in the sunlight, and the snakes would come out of the swamp, onto logs, to warm themselves in the afternoon heat.

Leaving our cars and trucks in the dirt parking lot of the Fakahatchee Strand State Preserve, we loaded two huge swamp buggies with drugs, gear, tent, tarps, inflatable bags, and radios and headed down an old railroad and logging spur. The buggies made a major statement: We were ready for nearly anything the swamp could throw at us. Their wheels were almost as tall as a man, and we lurched through holes on the spur big enough, it seemed, to swallow a Cadillac.

The Fakahatchee Strand, located not far from Big Cypress and the Florida Everglades and near Alligator Alley (a major road across south Florida), was a place of water and light. It did not so much exist, on that bright morning, as hover, all around us. As the water of south Florida flows toward the Gulf of Mexico, a vast river that moves so slowly it seems to be standing still, it cuts channels into the limestone that become drainage sloughs. The sloughs grow over with cypress trees, palms, and oaks into dense swamps that the locals call "strands." On that morning, the sunlight filled the shimmering air. The fluted trunks of cypress stood straight upright out of their own reflected images, and air plants draped through the air, living off the humid richness all around us. Insubstantial, full of reflected images and bright outlines, the Fakahatchee Strand seemed to exist somewhere between liquid and burning, dazzling air.

My first disillusionment came quickly: We weren't going to follow instinct into the strand. Though we would be using hounds to lead us to the panther, we weren't relying on backwoods wisdom, sifted through generations of southern folk. This was a high-tech adventure. We had radios, and the panther wore a collar in which there was a radio transmitter. Whatever secrets this cat had to teach, they weren't the secrets of the pristine swamp.

On the roaring buggies, we tracked her signal until we got close. Then we got ready to slosh into the swamp on foot.

The man leading this first stage of the expedition was Roy, a man whose hounds had a reputation throughout the south, Texas to Florida. He wore a white cowboy hat and, for swamp stomping, running shoes. He had a lean face, and he broke his taciturn demeanor only with terse comments. When I asked him what kinds of dogs he was leashing up, he shrugged, "Jes' hounds."

"They're great dogs," said the man in front of me, a biologist with the Florida Game and Freshwater Fish Commission. "Really well trained. They never bark unless they've got the scent."

Ready to go, we stood on the built-up edge of the spur, looking down into the water and the tangle of plants. I have to admit, I hesitated ever so slightly before plunging ahead, thinking of cottonmouths and alligators. I also thought of vaguer fears: of stepping in, of getting wet, of dark waters with no trails, of soft, muddy bottoms.

Of leeches.

Of getting lost.

We hedged, cheating on the last fear. Dave Maehr hung a radio collar on a tree so that no matter what happened, we could find our way out of the swamp.

I stepped down the dike and splashed into the water up to my knees. I was third in line, behind Roy and a biologist. Roy struggled to hold onto the dogs, who strained and pulled and tugged at their leashes. Nobody talked. Sometimes the water got shallow, covering only our ankles. We'd climb over hammocks—small dry rises in the swamp, covered with hardwoods like oak—and slog once more into the swamp water. We crossed running water, clear and waist-deep. Cypress knees bulged above the surface in the shadowy still places, and we stepped on them for firm footing.

The swamp was thick with wild coffee, sabal palms, bromeliads and epiphytes. Though the air sparkled with the bright sun slanting into the swamp, we still couldn't see much more than 20 feet ahead through the plants. Sharp fronds stuck at us, and underbrush slapped our thighs.

The biologist in front of me lost the sole of one of his shoes, and hiked the whole day on one bare foot in the mud and brush.

Every time I grabbed a branch for support, or touched a stick for balance, I looked for resting cottonmouths. I kept thinking of Marlow in Joseph Conrad's *Heart of Darkness*, going up the Congo.

For a while, it seemed that the cat was eluding us, moving farther off, and that we might not be able to catch up with her. We kept quiet to avoid scaring her, and we followed Roy, who was being pulled along by the dogs on their leashes.

After an hour, Roy gathered us into a group on a thick green hammock. The cat was just ahead. He intended to let the dogs go.

The dogs were churning on the leashes. When Roy set them free, they vanished into the thickets with a barkless explosion of hound-dog exuberance. They crashed through the trees, breaking, it seemed, every branch in their paths.

We stood on the hammock, hushed and hoping, the swamp suddenly still except for the racket from the loosened dogs.

When the hounds began a frenzy of yapping, whining, and barking, we knew they had a fresh scent. In the pure excitement of the chase, which lasted at most five minutes, the dogs kept up a chorus that was half howl, half bark, a high wail of canine desire and an ecstatic unleashing of animal energy. Then, as suddenly as they began, the howls grew slightly pitiful, throatier but thinner too, and we knew that the dogs had both succeeded and failed. The hounds were baying up a tree. They'd treed the panther, but *they* weren't going to get at her.

That's when all of us on the hammock broke for the tree with the dogs at the base. When we got there, we looked up to see, on a long, arching branch of a water oak, a burnt-yellow panther crouched on bent haunches, head crooked to stare down at us out of confused topaz eyes. She didn't look fierce. I'd expected a smoldering anger in her face: The mountain lion is a creature of elusive cunning, limber strength, hissing and spitting wildness. Instead, she looked passive, puzzled, a little wary, as if waiting for some clue about what to do next. Yet there were signs of power in her—in the torque of her crouch, in her supple flanks, in the breathtakingly long tail that flowed so beautifully behind her like smoke on a breeze. Her tail twitched back and forth with a nervous, curling energy—like the tic of a person who has repressed a great passion.

Her beautiful face was long and powerful and bold, in the high antique Roman style. There was black on either side of her nose and around her whiskers, which underscored the whole dramatic effect: superb aristocratic strength. Her quiet dignity was expressed in clean, sharp lines—at least, as much dignity as was possible at bay in a tree, surrounded by impossibly excited dogs and nine rubber-necking people.

We spent fifteen minutes looking at her. We exchanged a few comments, we smiled at each other, but mostly we were silent. One man said what a privilege it was to see this panther—he'd lived all his forty-five years in south Florida, and this was the first he'd seen. I knew how unusual it was to see a mountain lion in the wild, and the moments with this panther, before we got to work, were more than mere spectacle for each of us. In different ways that would have been hard to describe to each other, we each relished the moments privately, savoring the particular meaning of seeing this panther in this tree for each of our lives.

And then we did exactly what you'd expect Americans to do at a memorable moment: We all grabbed cameras and snapped pictures.

As we got to work, the story of this panther twisted in an unexpected direction, away from our brush with the sublime.

It really wasn't the indignities we made her suffer: Being shot with a tranquilizing dart, which dangled from her shoulder, decorated with red and white yarn like those prongs they stab into bulls at a bullfight. Slumping in a stupor on the branch and getting so tangled up that she was unable to drop to the inflated green pad below. Being lassoed by Jayde, who shinnied up the tree to lower her. Being slung like a limp load down to the ground. Being laid out drugged and comatose on the pad when we started to work on her.

Perhaps this kind of sophisticated intervention is necessary. I am resigned to it for highly endangered animals, and the controversy over handling panthers in Florida is similar to the one over the California condor. Whether to intervene with drugs and medicines, whether to capture panthers, whether to captive-breed them—these questions arouse strong feelings in people.

These indignities, however, were not what most affected me about this Florida panther.

III

If I made any rite of passage in that swamp, with that panther, it did not come in matching wits with a wily cat or in proving myself equal to the challenges of the terrain.

It came in seeing her close up.

She was sick, scrawny, and scarred. This female panther was between 9 and 11 years old. Under her left eye, she had a slashing wound, still open. The inside of one of her legs was cut. Her ribs rippled under auburn fur, which had no fat or meat underneath. Florida panthers are smaller than mountain lions in the western United States, for whom 80 pounds is a small female. An average female Florida panther weighs 75 pounds, and though this one weighed 72 pounds, she nevertheless looked emaciated. Ticks were everywhere on her, like so many black warts, bloated and fat with her blood.

By this time, Melody Roelke, the veterinarian, had taken charge of the operation, giving directions, assigning tasks, moving authoritatively.

"She's in bad shape," Melody said. "But still, she looks better than last year."

This panther represented the last of the cats that once occurred throughout the swamplands and wilds of the southern United States. Inexorably, they were shot by hunters and lost their habitat to agriculture and cities. Currently, the major known cause of death for the panthers in Florida is road kills. Venturing out of the swamps and onto highways, the cats get splattered by cars and trucks.

This female Florida panther and her twenty or so cosurvivors now live on about 1 percent of their original range. Melody thinks that the Florida panther has been forced to retreat into marginal habitat now, places that, if its habitat were more extensive, the panther would normally shun.

The fear was that the panther we treed might not, as a result, be healthy. The reason we caught her was to take blood samples, examine her reproductive system, and inject her with vitamins and minerals.

"In the strand, the panthers have low body weights and are anemic," Melody said. "Points to some sort of nutritional problem. Plus, they have heavy doses of parasites—ticks and tropical hookworms."

Melody asked me to record times and doses as she did the workup. The following is my contribution to science, and it may give you some idea of just what our rarest, wildest animals require in the way of help.

1:15 p.m.
Ketamine: 280 mg ("an excellent dosage," according to Melody)
Valium: 3 cc
Fluids: Lactated ringers (800 cc)
Heparin saline (intravenous)
Rompun: 3 mg
Aminoplex to I.V.: 40 cc

1:34 p.m.
Blood samples taken
Ticks: "100s and 100s and 100s" (according to Walter, who removed them with tweezers and dropped them into a vial)

1:40 p.m.
  Rompun: 8 mg

1:49 p.m.
  Ketamine: 100 mg
  Valium: 2 cc
  Tooth impressions: brown cast on right, red cast on left

2:04 p.m.
  Iron dextran: 4 cc
  Dronset: 4 cc
  Filobac: 4 cc
  Imrab: 1.8 cc
  Hypo B: 5 cc, plus 0.8 cc subcutaneous
  Ivomec: 0.8 cc subcutaneous
  LA 200: 4 cc
  ESE selenium: 0.8 cc
  Cranial left nipple: apparent old wound, slight thickening
  of skin under nipple, excretes a clear and golden fluid

2:33 p.m.
  Vagina: apparent warts or growth on inside; samples
  taken

2:44 p.m.
  Yohimbine: 0.4 cc

The yohimbine would help her regain consciousness more quickly. Some of the other drugs were worming medicines and feline vaccinations for various diseases. Before we finished the workup, Walter took off the panther's old radio collar and riveted a new one around her neck, with fluorescent green reflectors to make her shine in headlights on the highway at night. It might protect her from getting hit by a car or truck.

Even drugged and unconscious on her green inflated pad, covered with scars and infested with ticks and vaginal warts, the panther still had a powerful magnetism about her. All of us felt an amazing and powerful urge to touch her. I stroked her burnished fur, spread wide her massive paws (bigger than my hands), unsheathed her claws, felt the full, sensuous length of her exquisite tail. Touch is, I think, the most completely physical of the senses, even more than taste.

I was careful, too, to note the three marks that distinguish the Florida panther from all other mountain lions. First, she had a kink in the end of her tail, under the black tip, caused by the fusing of the last two vertebra. Second, she had flecks of white (perhaps from tick bites) in the otherwise auburn fur along her back. Third, there was a whirling cowlick of fur between her shoulder blades. Some biologists claim these features of the Florida panther are signs of a "genetic bottleneck," caused by severe inbreeding in a population of panthers so low that it may breed itself into sickness and oblivion. Others claim they are merely the marks of the race.

At 2:55 p.m., we put the panther into a tent which Melody had specially designed as a shelter for recovering panthers. We found the highest ground we could, to keep the cat dry. Melody did not simply let her go, or let us leave, because she did not want the panther to wander off before she was conscious enough to survive. If released before the drugs wore off, the panther could fall down, get wet and cold in the swamp, and develop hypothermia or drown.

That meant we had to stay in the swamp until night, since it would take hours for the effects of the drug to wear off. Most of the crew left. Only Jayde, Dave, Melody, and I stayed with the panther. We stood in ankle-deep water as the swamp grew dark.

Throughout the afternoon, I found myself increasingly impressed by Melody as well as the panther. Worried about the cat getting wet, Melody crawled into the tent several times as the cat was regaining consciousness. The panther would raise her head and make what sounded like alcoholic hisses and drunken, throaty threats. Undaunted, Melody would shift the cat to some drier spot in the tent.

Melody seemed to thrive on the cat's growing fierceness as she awoke. They were two tough alpha females: the panther waking, Melody caring for her in Levi's torn at the knees, long underwear for a shirt, and Converse tennis shoes.

Late in the afternoon, when Melody went into the tent, the

panther snarled and spit. With a paw much too fast to see, pure, invisible cat-speed, she struck out, swiped at Melody, and slashed her wrist. Melody enjoyed the cat's feistiness and liked being near it. About her cut wrist, Melody said simply, "Doesn't bother me."

By 5:30 the winter swamp was darkening, the color draining out of the royal palms and oaks and water—a screen of black before the last golden gasp of the sunset. At 5:55, in the graying light between dusk and full night, Melody decided to let the cat go. The tent was rigged so that we could pull ropes from behind to open the front panel and let the cat out.

We gathered behind the tent. Melody tugged at the ropes, and the front door opened. The panther refused to move. The tent was as good a place as any, I suppose, as dry as anywhere else in the swamp, to spend the night.

Melody tried to spook her, shaking the sides of the tent. We were no more than 5 feet from the panther, but I swear, we didn't see her go. In one extravagant, explosive charge, she simply vanished. One minute she was lying there, and we were debating whether she was ready to go, and the next the tent was empty—a motion, an exit, an escape beyond comprehension.

It was strange, her vanishing. We could hear her bounding through the underbrush, a few bolting strides with splashing sounds reaching us in her wake. But it was less like trying to watch something happen than watching something not happen—watching something that was suddenly not there. Through a kind of feline magic, the panther became instantly herself by disappearing for us.

IV

We were left standing in swamp water in the dark. Gathering our stuff, we headed east toward the swamp buggy. Only a few shapes stood out from the shadowy pall gathering in the swamp. The sky, lit by a full moon, hung above the swamp in the ultramarine blues of a Maxfield Parrish painting. The fronds of royal palms curved in dark profusion against the sky, arcing in lovely jagged elegance.

Dave guided us, following the beep of the radio receiver toward the swamp buggy. I would be lying if I said I wasn't nervous hiking in the Fakahatchee Strand at night, surrounded by dark trees and darker water. We just headed in the direction of the moon.

Of course we'd get out of here. Of course. We had the radio. We just concentrated on walking. Thigh-deep in the water. Hip-deep. Waist-deep. We walked for over an hour, longer than the hike coming in, but there was still no sign of the railroad tram we were trying to intersect. I kept a stupid watch for alligators gliding over the dark surface of the water.

For some reason, Dave kept angling us north of the moon; it seemed to me we were angling off course. Suddenly, without a word, Jayde broke from behind Dave and headed off in another direction. You tell me. You're in a swamp, and you have a choice: Ahead, a man with a radio receiver, but a newcomer to this swamp; to the left, a man acting on instinct, but a veteran of the swamps. Which one would you follow?

I sloshed off behind Jayde.

It took him only ten more minutes to get us out, though the final stretch of water was another deep one, waist-deep. My heart jumped into my throat when I saw, just 2 feet away from me, what I was sure was a floating alligator. Cautiously and nervously, I bent down for a closer look. It was just a rough log.

Scrambling up onto the railroad spur under a moon-rich sky, we squished in wet tennis shoes toward the buggy. Once the stress—never very serious—of getting out of the swamp was passed, I found myself focused on the panther's green reflector collar. Though we still try to see the wildlife in America as untrammeled, there is little left that is pristine. This is no exaggeration: The wilder and more spectacular a creature is, the greater the likelihood in America that it is tagged or radio-collared, even surviving on dosages of medicine. Few wild animals are seen anymore except by the biologists who make it their living to chart—and save—these creatures' lives with all the paraphernalia of high technology. The reality is that much of our wildlife has been lost and most of what is left wears collars.

It's a serious question: Are we simply going to trust in the wizardry of science and technology to save these creatures, and run the risk of somehow disowning them ourselves, of abdicating our responsibilities for their plights? A captive-breeding program is probably inevitable for Florida panthers—it is already on the table. Captive-breeding programs are seductive: They produce young, are glamorous, and get publicity, but they treat only the symptoms of the loss of species.

Will the panther survive? It's hard to say. The species has major problems. Deer hunters have tried to deny this, and they have resisted the move to protect panthers, fearing that more panthers will mean fewer deer. In a hopeful development, the state recently allocated huge amounts of money to build wildlife underpasses at major highways in an attempt to reduce road kills. It may be, though, that too much Florida habitat has already been lost because of population growth—what remains may not support many more panthers in the wild.

Since that day in the swamp, Dave Maehr has examined sixteen other Florida panthers. He has come to believe that reproduction is taking place at replacement levels and that for this small population the crisis stems from "uncontrolled and unmitigated habitat loss."

The female we treed and treated in the Fakahatchee Strand was captured and removed from the wild on April 13, 1987. She had been having no babies in the wild, so Melody and others decided to try to breed her in captivity. When they captured her, she was in terrible shape. She was old and had lost more weight since we had treed her. In captivity, she gained some weight for a while and seemed to be responding at first. In 1988, she was moved to a reproductive location with seven captive mountain lions from Texas.

The other cats responded to mates, but she did nothing. She was menopausal and senescent, and she languished away. In August 1988, she died of renal failure.

FIVE

SIRENS

I

Cheap-TV skies—black and white and blurred, oppressive. Not the rich pastels and fiery magentas of the Florida of our fantasies. Clouds like dark etchings, hanging over the wide waters of the river, dripping as if they'd just been lifted out of a dark, acid emulsion. The Crystal River—the meaning of its name lost in the obscure, metallic dun of its winter-gray surface.

Dressed in a diver's black wetsuit, cheeks squished beneath a face mask, lips bright pink with the cold, I stood in a boat, scowling at another boat of divers. The morning was corrosive, and I was feeling murky and mean.

Every year, 65,000 people come to King's Bay, the location of the Crystal River National Wildlife Refuge and manatee sanctuaries, to swim with the manatees. It's a big tourist draw, just 7 miles up the river from the Gulf of Mexico. A flat-bottomed barge was chugging out to the fluorescent orange floats, so gaudy on that dreary morning. There it would tie up and disgorge the crowd of laughing and giggling people for their innocent fun. If you're optimistic in outlook, more mainstream and mellow in your opinions, the scene could be proof to you that manatees and people can live together.

But if you're inclined to anger, it looks more like harassment than fun.

Yet there I was, too, diving with manatees.

Florida is supposed to be the new American playground—a peninsula in the Sun Belt's perpetual paradise. The Edenic dream of every senior citizen looking for eternal youth, it is the New World's new world: a fruitful and fulfilling garden of oranges and sunshine. No terminal disasters. No time unless it is ripening.

Florida is not so much a place as a fantasy. A cultural topos. Florida is the future of a nation heading for a secular, hedonistic, material perfection. History and death vanish there under the charm of our imagining.

Close—but not quite.

Florida: Behold our future. The state ranks third in the nation in the number of endangered species already listed, and that number is probably climbing. As of 1986, 530 animals and plants were included on the state's endangered and potentially endangered lists, with 419 on or under review for federal lists. Third behind two other sunshine states, Hawaii and California. Hawaii, the southwest, and the southeast have accounted for 92 percent of the total extinctions in the United States as of 1977.

Not a perfect fantasy after all.

Endangered species have become part of our national conscience: They remind us that our lives and our pleasures are often purchased at the animals' expense, an invasion of their territory. It's part of the anger, sorrow, and guilt that everyone I know who works with endangered species feels at times—the sense that we're all accomplices in the problem of endangered species, because of the way we live.

I slid off the transom of the boat and lowered myself into the water, not far from a bright orange buoy. The warm water seeped under my wetsuit and I floated, facedown, looking for manatees. It didn't take long to find one, creeping along the bright limestone bottom about 10 feet below me like a gray shadow. She started to rise toward the surface, as if curious, as if she had found me instead of the other way around.

Sculling with her stubby flippers, she rotated indolently as she rose, bringing her immense heft and girth around to face me. She was about 8 feet long and weighed about 1500 pounds. With a tinge of panic, a hint of terror even, I braced myself for the encounter. Manatees are supposed to be gentle, but you never know.

Once at the surface, she came at me like a huge, vague idea taking shape. She looked something like a walrus, with a big flat nose bristling with hairs and leathery skin that resembled cracked pottery. Her flippers were short, and I could see the bones moving under their skin like fingers inside a child's mittens. Sunk deep into her face, her pinprick eyes seemed much too small for her bulk, and the cloud of bleary film across them gave her a drugged, mindless expression. But she had a mind of her own, and kept coming toward me in a direct, guileless approach.

Then, with a jolt, her snout banged gently, with sudden intimacy, into my face mask, as surprising as a first kiss.

My earlier encounters with endangered animals had been intimate, but not quite like this. In the wolf den, in the sparrow's cage, in the air with condors, in the swamp with the panther—I had always been just a spectator, swaying between a world I didn't fully belong to and a self I didn't fully know.

Here, however, I was in the manatee's element, a participant with her in a slow-motion drama. There could be no illusion in this one that I was the author of what was happening.

We drifted in a slow circle, locked mask-to-nose in a mutual gaze. Something about her effortless swimming, her massive presence so at ease in the water, made me feel awkward and nervous. Through my face mask, I had only a small circumference of vision, as if I were looking through a tube, but I didn't want her out of my sight.

She drifted lazily to my left, and I found myself sculling frenetically to keep her in view. I had to see her to feel safe. I bent in contortions at the waist to face her, kicking stupidly with my flippers. Unable to keep up with her, I started to lose sight of her as she slid around to my back. I did not want her behind me.

Something about her graceful drift made me feel slightly

ashamed—for all the moves I use to try to get ahead, for all the hustle I go through just to keep up with my life.

That's when I realized there was no point trying to keep her in sight. I wanted to keep her under some kind of control by seeing her; I suddenly felt tyrannized by my need to see.

Though I had not realized it when I went to Crystal River, I had intended to learn *about* manatees. It had not occurred to me that another posture was possible, that I could actually learn *from* manatees. My sense of superiority to nature is so deeply ingrained that I feel most comfortable in watching the world out there, in spying and in staring. I also normally prefer the kinds of authority and power that are expressed in metaphors of vision—when my ideas are clear, when I see things well, when I am lucid. The eyes are the instruments of reason: We speak, for example, of the Age of Reason as the Enlightenment. Yet Hans Jonas, in *The Phenomenon of Life*, shows the implications of this modern dependence on vision in knowing the world: "Thus vision secures that standing back from the aggressiveness of the world which frees for observation and opens a horizon for elective attention. But it does so at the price of a becalmed abstract reality denuded of its raw power."

For me, needing to see and observe kept the manatee from existing except as she was seen. In our clear-eyed, Galilean age, we are dominated by the knowledge that comes from the eyes. It keeps the world apart, known only on the surface, understood by appearances. It makes the world anonymous, and however much it gives us, it also makes us strangers.

The manatee at Crystal River embarrassed me out of my superiority. What's missing when we try to control the world through our eyes is genuine reciprocity, and with that loss we also lose the kind of knowledge that comes not from explaining but from participating. Greater intimacy is possible, but it means abandoning our supposed superiority: Rather than knowing some absolute truth about an animal, we enter into dialogue; rather than extracting some fact about an animal, we make that fact live with significance in our lives. Greater intimacy also means recognizing our own role in shaping and creating what we know. Every way of knowing implies a kind of relationship with the world. We have faith in facts for their own sake, the desire for absolute truths, because they offer security and a sense of control. On the other hand, dialogue implies change and uncertainty. But, in place of hard knowledge, it offers understanding and the possibility of wisdom.

I quit paddling so nervously, quit the hyperactive sculling. I relaxed under this awful burden of self-consciousness I carry around with me, from too much education, from being smart but never smart enough. Yet something was here for me, and I didn't have to do anything but let it happen. I quit struggling to watch her, quit worrying about where she was. I tried to understand this manatee, and this moment, with parts of me that I often can't find, parts that are lost or withered—with a quality that makes rationalists nervous: empathy.

The manatee moved around me in slow circles, and I just drifted with her. She floated behind me, and I felt her bumping my back with the tender thuds of 1500 pounds.

She seemed to love to touch. She dived below me, came up, and banged into my belly. She rolled over, sinking; then, like a huge slowly spinning bubble, she bobbed toward me and nudged my side. Wafting her flat round tail, she piked toward the bottom, where she tumbled onto her back with an effortless and elephantine magnificence.

Then, plumping back to the surface, she came at me again with her infinitely engaging, unpretentious bulk, moving in from an angle, keeping me in her sideways glance. We drifted closer, her approach eloquent with curiosity, playfulness, and an unbelievably endearing vulnerability. The way she moved, somnolent and superbly languid, I almost felt as though I were napping in the warm waters and she were a creature coming out of the dormant margins of my mind.

She touched her bristled snout to my face mask, and we floated in a balletic swirl. Manatees may be ugly, but I saw in her all the serene beauty of things that simply are what they are.

I was honored by her trust, no different from the honor I feel when another person reveals a part of himself or herself to me and trusts an impromptu, naked moment. With the manatee, I pushed back some boundaries, felt a little larger. Because under all my reasons, all my spasms of haste and dispatch, all my aggressive posturing, she made me recognize in her the one thing I struggle most to admit in myself—an essential vulnerability.

It was just this vulnerability in the manatee that I would find violated later, at Blue Spring State Park.

II

It's an unlikely association: The manatee belongs to the order of Sirenia—sirens. It is one of four surviving members of the order and is technically called the West Indian manatee with the scientific name *Trichechus manatus*. The others are the Amazonian manatee of South America, the West African manatee, and the dugong of Australia and the Indian Ocean—all tropical and subtropical. A fifth species, which grew to almost 10 meters in length, was the Steller's sea cow. Georg Wilhelm Steller discovered it in 1741, on his voyage to Alaska with Vitus Bering, and gave us the only firsthand account of this northern sea cow. It went extinct a few decades later, presumably through overhunting by sealers and whalers.

Sirens. They were the birdlike women who for Homer's Ulysses symbolized the temptations of sensual pleasure. By the Middle Ages, the mythological sirens had become confused with fishlike mermaids, who lulled men through their music into a sleep that destroyed them.

How could it be that humans might associate a manatee with erotic and imaginary sirens?

No less a figure than Christopher Columbus made such a mistake when he saw manatees for the first time. On his first voyage to the Indies, when he discovered the New World, Columbus made what is now considered the first sighting of a manatee by a Westerner. On Wednesday, January 9, 1493, he was coasting the shores of Haiti. He wrote in his journal of sighting three sirens, which he called "mermaids." They rose very high out of the sea. He wrote, "They were not nearly as beautiful as painted, although to some extent they have a human appearance in the face."

Perhaps these weren't manatees. But if they were, Columbus

could have had no language for his experience, no concept for the creatures he saw. Here is a glimpse of an explorer confronted with something genuinely new. In some way, he couldn't see them—at least not as new creatures in a new world. Instead, he assimilated them into the Old World mythology he carried with him in his head, filtering the reality of his experience through the world view he inherited. This is an example of the politics of perception, a composite of observation and desire, a mix of reality and dreams. In metamorphosing real manatees into imaginary sirens, Columbus illustrates how difficult it is to locate ourselves in the world, to distinguish between the world outside and the world inside.

It must have made great sense to Columbus that he might see sirens, since he was on a heroic quest, a great traveler like Ulysses. Perception is not simply passive; it assimilates and transforms what is observed. In order to see, we have to make the world over, make it ours.

In Columbus's time, myth and nature were not so clearly separated as they now are. Using the image of the mermaid, Shakespeare dramatized the association between nature, imagination, and dreams in one of his most beautiful and most popular plays, *A Midsummer Night's Dream*. In this play, written late in the Renaissance, on the brink of the Age of Reason, Shakespeare presents a view of nature that was soon to be lost in a more enlightened and scientific world. In the famous speech by Theseus, Duke of Athens, the imagination is described as discovering the "form of things unknown." The poet, Theseus says, not only perceives these forms but speaks in "fancies images," which is more than "cool reason ever comprehends."

These images are not true in a factual sense, but they are "something of great constancy"—true in a higher sense, as Shakespeare says. There is a poetry in nature that only the imagination can respond to or perceive. One lovely passage uses the image of the mermaid to represent this other nature:

> *Thou rememb'rest*
> *Since once I sat upon a promontory,*
> *And heard a mermaid on a dolphin's back*

Uttering such dulcet and harmonious breath
That the rude sea grew civil at her song,
And certain stars shot out of their spheres
To hear the sea-maids music?

The allusion to the music of the spheres—stars shooting out of the sky at the sea-maids' music—is a reminder that the cultivated soul, the well-tuned imagination, can hear this other realm, which is associated with nature.

The split between reason and imagination widened in the century after Shakespeare. Science and reason gained power, and the imagination was reduced to the decorative illusions of the baroque.

In *Man and the Natural World*, Keith Thomas documents the "revolution in the perception" of nature that took place during this time. Before the Renaissance, the world was understood largely in terms of its parallels with human culture—morality and politics, mainly, though also psychology, or human character. In explaining how the new naturalists of the period changed the way we see nature, Thomas writes:

> But in eroding the old vocabulary, with its rich symbolic meanings, the naturalists had completed their onslaught on the long established notion that nature was responsive to human affairs. This was the most important and most destructive way in which they shattered the assumptions of the past. In place of a natural world redolent with human analogy and symbolic meaning, and sensitive to man's behaviour, they constructed a detached natural scene to be viewed and studied by the observer from the outside, as if by peering through a window, in the secure knowledge that the objects of contemplation inhabited a separate realm, offering no omens or signs, without meaning or significance.

I have shown some of this premodern thinking as it appeared in the bestiaries of the Middle Ages, where wolves and condors offered images of human life. It appeared also in Christian thinking, in which the world was seen as an encoding of signs and signatures, teaching God's design as well as correspondences between the spiritual and the natural worlds.

In *The Order of Things*, Michel Foucault calls this an "analogical cosmography." Knowledge was constituted through resemblance and analogy. That is how people *knew*. In a world "saturated with analogies," Foucault says, a man "stands in proportion to the heavens, just as he does to animals and plants, and as he does also to earth, to metals, to stalactites or storms." With the emergence of rationalism, as Foucault calls it, in the seventeenth century, knowing by analogy was displaced by analysis:

> All this was of the greatest consequence to Western thought. Resemblance, which had for long been the fundamental category of knowledge—both the form and the content of what we know—became dissociated in an analysis based on terms of identity and difference. . . . As a result, the entire *episteme* of Western culture found its fundamental arrangements modified.

So in Oberon's apostrophe to mermaids singing on a dolphin's back, we are being called to participate in this now-vanished vision of nature. The sirens are in fact symbols of nature as viewed through the human imagination—a nature alive with metaphor and resemblance: of half-women and half-fish, of songs that make seas civil, of stars that respond to music, of the transformation of nature into a harmonious vision, of the strange and wonderful.

Finally, T. S. Eliot uses the mermaid image to describe the spiritual and imaginative desolation of the twentieth century. In the famous passage from "The Love Song of J. Alfred Prufrock," Eliot has Prufrock say:

> *I have heard the mermaids singing, each to each.*
>
> *I do not think that they will sing to me.*
>
> *I have seen them riding seaward on the waves*
> *Combing the white hair of the waves blown back*
> *When the wind blows the water white and black.*

We have lingered in the chambers of the sea
By sea-girls wreathed with seaweed red and brown
Till human voices wake us, and we drown.

The mermaids suggest Prufrock's alienation from himself and life. The fear for our time is that we are drowning in the sprawling wasteland of civilization. Noah's ark may be one image for the modern attempt to save endangered wildlife, but the imagination offers another ark, where animals can be reborn inside of us. If we can never know what an animal feels, if we can never know the world of its consciousness, if, in short, we can never know what it is like to be other than we are, we can at least imagine it.

III

Even on an unusually cool winter morning, Blue Spring had a tropical feel to it. Lining the shores, crowded against the stream banks, the trunks of sabal palmettos rose up into their plumed effusion of leaves. A few saw palmettos here and there, leaves spreading outward, green fireworks on a dark day. Spanish moss dripped off trees in a kind of viscous ooze all their own, and winter-brown buttonbrush tangled along the ground. Dark clouds rolled through the sky, slightly threatening, but the slack drift of the waters had a hypnotic, almost narcotic tranquility. Steam played over the surface of the cloud-darkened run, and under the diaphanous sheen of the emerald-pure waters, bright sands lay like sheets in the streambed.

Blue Spring State Park, just north of Orlando, is a well-known wintering place for manatees. In fact, the colonial American naturalist John Bartram first found manatees wintering here in 1766, on his explorations of Florida. In a clear flow from the depths, spring-water boils up out of caverns and caves and lapses into a gentle "run" (as it's called) for several hundred yards to a dark confluence with the cooler, swamp-bordered waters of the St. John's River. At about 72°F, the waters of the run at Blue Spring form a winter haven for about seventy manatees, protecting them from hypothermia, which can be and has been fatal for them in cold spells. The current census of manatees in the United States puts their number at about 1200, and though the entire state of Florida has been declared a sanctuary for these peaceful animals, they have only six natural warm-water springs where they can congregate during the winter.

About 200 yards from the confluence of stream and river, seven manatees dozed along the bottom, their long noses anchored to the sand, their bodies forming a fat arch toward their rounded, manhole cover of a tail—humpbacked and hunkered on the bottom.

I was with a man named Wayne Hartley, a park ranger for eight years who had been gathering data on the manatees of Blue Spring for the Sirenia Project in Florida, managed by the U.S. Fish and Wildlife Service out of Gainesville. Wayne cared deeply about the manatees; he knew these animals in a personal way.

We watched the sleeping manatees from a large hammock beside the stream. I crawled out onto the trunk of a large oak, leaning over the water, where I could look straight down into the stream. Every fifteen minutes or so, the manatees would slowly float to the surface, their heads tilting up first, with their tails drooping behind. They rose in unison, first one and then another lifting lazily off the bottom in a stylized aquatic choreography, their buoyancy provided by huge lungs. Breaking the serene surface of the water, their snouts sent up a swoosh and a small feather spray. Then the manatees settled sleepily back to the bottom, to the exact spot they had occupied before they rose.

There was something totally imperturbable about them as they rested and breathed. According to Wayne, the current speculation is that manatees don't even have to break their sleep to breathe. They sleep right through the whole slow motion of rising and breathing and returning. Watching them, you do get the feeling that they excel at sleeping, that it's one of their specialties, so to speak. I couldn't help wondering what dreams, flitting like fish through the seas of their sleep, curled their lips in such self-satisfied smiles.

Their smiles, however, have nothing to do with emotions but are the result of anatomy.

The manatees' flippers are interesting. Short, stubby things that droop down from their sides in little arcs. The flippers do not look

A small, dark bird, the dusky seaside sparrow foiled even the most ingenious attempts to save it. A decade after the last five birds were captured in the wild, the species became extinct in 1989. *(Photo by Paul Sykes, Jr., United States Fish and Wildlife Service)*

Puerto Rican parrots preen themselves and each other. Their bold white eye ring seems to exaggerate their vigilant and searching gaze. Intensive and ongoing efforts by biologists in the Caribbean National Forest were needed to bring the parrot to its current precarious recovery. *(Photo courtesy of the Caribbean National Forest)*

Once distributed throughout the Southeast, the Florida panther is now a very rare creature, confined to a few swamps in southern Florida. *(Photo by Dave Maehr, Florida Game and Freshwater Fish Commission)*

In the West, the mountain lion is reduced in numbers and persecuted, but retains its incomparably dignified profile. *(Photo by Art Wolfe)*

California condors are now extinct in the wild. They once roosted, like this one, on the largest branches high in conifers.
(Photo by Helen Snyder)

Biologists such as Peter Bloom of the National Audubon Society captured the last of the condors in the wild to radio tag them and check their blood for poisonous lead. Condors were one of the largest birds in the Western Hemisphere.
(Photo by Helen Snyder)

Despite its name, the most distinguishing characteristic of the black-footed ferret is its black face mask. It is a member of the weasel family. *(Photo by LuRay Parker)*

Black-footed ferrets may now be extinct in the wild. They lived in the burrows of their favorite prey, prairie dogs, throughout the Great Plains. Only two colonies of ferrets have been studied. *(Photo by LuRay Parker)*

With its leathery skin, pinprick eyes, and huge, bristled snout, the Florida manatee lives in the warm waters off the coast of Florida, where it is often run over by boats and barges. *(Photo by John F. Wahl)*

Manatees seem to enjoy touching and have a serene vulnerability about them. Note the nails on the manatees' flippers, vestigial signs of their closest modern relative, the elephant. *(Photo by John F. Wahl)*

The recovery of the whooping crane, after half a century of conservation, is a fragile triumph. *(Photo courtesy of Tom Stehn, United States Fish and Wildlife Service)*

Above: A North Atlantic right whale rolls into a deep dive, its broad flukes like a banner above a calm sea. *Below:* A mammoth creature, this right whale may weigh forty-five tons, and seems to defy the elements with its breaching in the Bay of Fundy. *(Photos by Gregory Stone)*

Very few places in the lower United States now support the gray wolf. It is both an emblem of wilderness and the object of controversial "wolf-control programs." *(Photo by Art Wolfe)*

Gray wolf pups are usually born in early spring in an underground den, abandoned beaver lodge, or hollow log. *(Photo by Art Wolfe)*

The northern spotted owl sits in the center of intense controversy over the loss of the ancient forests in the Pacific Northwest. *(Photo by Art Wolfe)*

This family of trumpeter swans on the Kenai Peninsula in Alaska is part of an exhilarating recovery for the species along the West Coast. *(Photo by Art Wolfe)*

like those of, say, a seal. They are stiffer, more like truncated arms than flippers, and the manatees often use these appendages to help them crawl along the bottom of a stream. On the flippers, at the very ends, are three or four white "nails." Fingernails.

These fingernails are vestiges of the manatee's origins. The closest modern relatives to the manatee are not any of the creatures of the sea. They are, instead, the elephant and a small mammal called the hyrax, a rabbitlike animal of South Africa. The vestigial nails on the manatee's flippers are descended from the foot of an elephant. Manatees also have vestigial pelvic bones, as well as similarities in their blood to the elephant. In a kind of extravagant rebuttal of the general course of evolution, which we normally conceive of as a grand progression from water to land, from fish to humans, the early Sirenia went the other way—marching, as it were, against the current. The Sirenia may have been the first group of mammals to return to the sea. Over 60 million years ago, the ancestors of the manatee traded land for water, evolving into aquatic mammals. Even the manatee's nose, with its prehensile lips for gripping food, are reminders of its relation to the elephant, and its unhurried swim through the epochs.

Almost everything we have learned about manatees appeals to that part of us that loves a peaceful, unhurried approach to life. They migrate to warm waters in winter, they have no enemies (except people), and they are harmless vegetarians. In fact, Sirenia are the only marine mammals that are vegetarians. Manatees spend most of their time eating (when not sleeping). Grazing their underwater pastures, they can eat 100 pounds of plants in a day, stuffing their favorite food, water hyacinth, into their mouths with their flippers.

Essentially solitary animals during the summer, in the winter they sleep, swim, and play in groups. They swim in semisynchronized formations, abreast or in single file, moving up and down the run, making maybe 2 to 4 miles per hour, ambling like trucks with the governor on low. Some of them even crawl along the bottom, and if you look carefully through the water, you can see where they've dragged their flippers and tails, a series of streaks and semicircles writ like fragile characters into the sand on the bottom.

One of the recent discoveries, the result of the sort of research

Wayne has been conducting as he watches the manatees every winter, is that the bond between mothers and babies seems to be stronger and more durable than any other in the species. A mother gives birth to a 65-pound, 3-foot calf every two or three years, and the calf and mother stay together, winter and summer, for up to two years. The little calves are as adorable as stuffed animals, swimming close to mother and suckling from nipples tucked under the mother's flippers.

A female can come into heat at any time of the year, and she will attract a herd of breeding males. But even in this sexual situation, manatees show little aggression. At most, the males just bump and shove each other. They seem genuinely to enjoy contact and to love playing. They roll around each other and even embrace with their flippers, touching noses as if they're kissing. They will sometimes move into the shallow waters at Blue Spring and start thrashing about in exuberant play. Tails thrown out of the water. Spray, splashes, and foam. They'll even lift calves completely out of the water.

The fine hairs on their skin help them feel things, and the bristles on their prehensile noses are part of their olfactory system. Everything they come in contact with, they smell and touch. This is how they identify their world.

Manatees do not fight over food, nor do they defend territories. In the summer, when they disperse from their winter refuges, they swim extensively, in a range that can take them up the East Coast into North Carolina and along the Gulf Coast to Alabama and Mississippi. The majority of them, however, stay in Florida year-round. They're strong swimmers, though it's hard to believe, watching them lounge around in Blue Spring, that they can get around so much. But radiotelemetry studies are showing just how far these manatees can go on their slow cruises.

Because they have no natural enemies, they have no defenses, except for an occasional quick burst of speed, up to 15 miles per hour, when frightened. And this is the most endearing part of the manatee's charm—its complete vulnerability.

Though protected by their tough skin and a thick layer of blubber, these roly-poly, pudgy creatures have a low metabolism that leaves them vulnerable to cold. Cold winters can kill them. The Christmas freeze of 1983 led to a wave of manatee deaths from hypothermia.

But what is saddest about the lives of manatees, sadder than their vulnerability to cold, is the pain they must suffer from their conflicts with humans. The number of people moving to Florida every year, for the sun and water, is staggering, and it suggests the trouble ahead for wildlife in the state. Between 1987 and 1988, in one year, the net growth in population in Florida was 373,998, according to the state census figures. The year is typical of recent trends, and over 80 percent of the growth is expected to occur on or near the two coasts. That means new houses on the shorelines, new marinas, and more boats. The manatee has been, and is, on a crash course with the increasingly marine lifestyle of Florida. The loss of shoreline habitat is indirectly squeezing out the manatees.

The manatees also face violent suffering. About 125 manatees die each year in Florida, and about 30 percent of those deaths are caused by humans. In 1985, for example, thirty-nine manatees died of human-related causes. Most of those deaths fall into one category: collisions with boats and barges. In recent years, the numbers seem to be getting worse. In 1987, for example, thirty-nine manatees died from collisions with boats and barges alone.

About 80 percent of all manatees are scarred by propellers. Ironically, researchers identify particular manatees by the scars on their bodies and tails and flippers. Of the 1200 or so manatees, over 850 of them have been individually identified. Every year, several manatees suffer such terrible wounds from collisions with boats that they have to be removed from the water and taken to, say, the Miami Seaquarium for treatment.

On a trip to the Miami Seaquarium once, I saw a manatee that had just been brought in from West Palm Beach. Its entire tail had been chewed off by the propeller of a boat, and before she was found, the open wounds had started to rot. Whether she would live was still a question. Many like her die of pneumonia.

Boats cause immense suffering and pain for the manatees. Many of the manatees are simply crushed. They may show no external wounds, but their ribs break and puncture the lungs, pierce the diaphragm, rupture the intestinal tract. Or the props tear their flesh. Make no mistake. Manatees may look like dumb brutes, but when they're hit, they show all the signs of pain as they thrash and twist in the water.

The agonies the manatees must endure came home to me when Wayne and I got into a canoe to conduct his morning census of the manatees at Blue Spring. As a ranger at the state park, Wayne had charted most of the manatees through the years, watching mothers return with their calves. Wayne thinks that boaters typically don't realize they pose a danger to manatees and that even after they've hit one, they think it was probably an underwater log. They are simply unaware—some people even drive their boats when they're drunk.

Big boats like barges that crush manatees may have no idea they've even run over something. Pleasure boats are getting bigger, too, and more powerful, such as the cigarette boats capable of 90 miles per hour.

"Part of it," Wayne said, "is that the boats are so fast. And the manatees are so slow. The manatees don't know how to deal with them, and they have such good hearing, they appear to get confused, don't know what to do."

We paddled slowly up the run toward the springs. The sun broke tentatively through the clouds, and I dangled my hand into the water, which felt warm as blood. In an oak, a barred owl watched us watching the manatees. Along the bank lay an alligator trying to catch some of the fugitive sun. We trailed the manatees, who, roused from sleep, had started their own parade into the current, single file, moving slowly.

All the manatees we saw that day were identifiable by the scars they wore from collisions with propellers:

Phoebe and her calf: Phoebe had a gash down her back and three nicks in her tail; half her left flipper was missing.
Success: Huge ugly gashes ran down her right side. The year before, two of her ribs had been poking out of the wounds. Now the scars were just two white brands: one, like a jagged

tear down her side; the other, evenly spaced slashes from the propeller.

Vasco: Her back had been run over by a boat. The propeller left a series of scars across her.

No Tail: Two-thirds of her tail had been sliced off.

Big Emma and her calf: An old mother, Big Emma had four perfectly symmetrical, straight-line slashes like ladder rungs down her right side. Her young calf, only 4 months old, swam beside her.

We saw twenty-seven manatees that morning as we canoed the run, each one of them scarred. They just kept passing below us, their backs covered with scars from our machines. Frightening scars, because they were so regular: laid out upon the manatees in mathematical precision, an artificial geometry of pain.

I felt for those manatees in an almost personal way—the way the grief of one heart implicates the grief of another. I was moved by the way they wore their identities, at least to us, so visibly in their wounds. It made me angry, too, at a culture that can inflict so much damage on creatures and not even be aware of what it's done.

I was reminded of a dream: I was walking along a freeway on a ridge. Cars were speeding past me, and I was confused. Suddenly, I was on a suspension bridge, and more cars were zooming past me with frightening speed. Somehow, I found a doorway on the bridge, and, entering, discovered an escalator going down. I stepped on and went down, one floor, two floors, three floors. Getting off at the bottom, I opened another door. In the room, I was suddenly naked in a group of naked people. All of us had wounds and lacerations.

Beneath all our speed and power, I thought, everybody's like a manatee. Inside our metal cars, which we slip into as though they were armor, we are all naked and wounded.

IV

Still, the manatees offer a quiet, calming sense of hope for endangered species. The hope for manatees comes from a number of sources, but it is personified by Jimmy Buffett, the southern songwriter who also chairs the Save the Manatee Club. I met him at his hotel in Coral Gables, Miami, where he was stopping over on his way from his home in Key West to Louisiana. Paneled in oak, the hotel room was the picture of casual southern elegance, and Buffett himself was a Biloxi boy who had found the good life—he was a celebrity. Tanned, wearing a gold chain around his neck, he'd just gotten in from fishing off the Keys.

Friendly and unaffected, he said he loved manatees because they remind him of Mark Twain, because they actually look like Mark Twain. "If Mark Twain came back to life, I think he might come back as a manatee," Buffett said with a flash of his burlesque humor. "I think the manatee is representative of the lifestyle in Florida. That's how I see 'em—as something to be preserved. That's why we're all here, to live a casual life and soak up some good ol' sunshine. They just want to swim in warm water, eat, fuck, and be left alone."

He talked on, with abundant energy, in sentences that would be nearly impossible to punctuate. He'd been working on a project to get Disney productions to invent manatee cartoon characters named Bubba and Hyacinth. He wanted to make the manatee a household word.

What struck me was that the manatee has a chance—maybe not for huge numbers to live but certainly for the species to survive beyond immediate danger of extinction. The association between Jimmy Buffett and the manatee suggests the kind of cause célèbre that the manatee has become in Florida, the kind of energy and resources that have been galvanized on its behalf. A good friend of mine, Judith Delaney, has performed heroic work in organizing and running the Save the Manatee Club. She works hard and tries to stay optimistic, stressing what has always been one of the main weapons in an environmentalist's arsenal: public education. The club had 800 members in 1984, and by 1988, Judith had built that number up to an astounding 16,000. The club supports fund-raising, teacher education, and a program for adopting manatees.

Biologists and state agencies are also working to better protect the manatee—requiring licenses for boaters, posting manatee areas for lower boating speeds, and instituting a policy that would protect state shorelines through a formula for the spacing of docks, boat slips, and marinas. And Florida Power and Light has proved very protective of the manatees that have taken to wintering in the effluent-warmed water around its more than two dozen power plants.

The manatee is a case study in hope for endangered species. Its problems have been recognized before its population has dropped desperately low, biological resources have been devoted to studying its life and protecting its coastal habitats, legislation has been proposed on its behalf, and the public has been educated for its support. The manatee is a reflection of what can be done for a species, given resources and dedicated people.

Still, a number of matters concerning the recovery of the manatee, and endangered animals in general, trouble me. Somehow, even with optimistic case studies like the manatee, we don't address the underlying problems of population growth and habitat loss. And we never get at the deeper roots of the problem—the Western mentality. Instead, we continue to treat animals as if they're separate from and external to us.

Meanwhile, the problems with the environment keep getting worse. Yes, we save chunks of land, but these are mere gestures, like beautiful flags in a gale. They're trade-offs, really, for the loss of even larger chunks. It's amelioration. Americans are aware of the problems of endangered species, and public-opinion polls show that Americans support environmental protection. We value wild animals, the polls say. Yet, all the while, the economic and political forces that drive our country continue to hold sway. Florida grows by over 1000 people per day. That means swamps are drained for land and water is diverted for the daily shower. Once, there were 2.5 million wading birds in the waters of south Florida. Now, that number is down to 150,000 to 200,000.

It's not hard to conclude that the problem of endangered species has not sunk very deeply into the national consciousness. Do we mean what we say or what we do? We go through periodic spasms of remorse, designate people to write reports, and barbecue another chicken in the backyard.

What's most worrisome is that environmentalists may unwittingly be part of the problem, co-opted even as we fight for guerrilla victories, swallowed by the larger forces at work. What, after all, is the liberal environmentalist's dream? He thinks 5 more miles to the gallon will save the planet. He believes in technology, because it gives him all those low-impact toys from REI and L. L. Bean. He talks seriously about such politically loaded concepts as "minimum sustainable populations," wondering how many manatees might be enough. He helps the bureaucrats divvy up the earth so that we'll slow the rate of destruction.

After I met with Jimmy Buffett, I went to the Florida Keys and met someone who gave me a different perspective on the American approach to life. Bridgett was blonde, with a face round as the full moon. We met on Key Largo, at a pond along the highway, where we had both pulled over to look at endangered roseate spoonbills—beautiful pink birds with bizarre spatulate beaks.

Bridgett had been educated at Paris; she was a Sorbonne Ph.D. in fact. Right away, she told hysterical stories about famous French philosophers—Jacques Lacan, especially—chasing after beautiful undergraduates on long weekends in Venice.

She loved birds.

In ironically flamboyant tones, I told her I would show her a magnificent frigatebird, soaring high above a blue tropical ocean. I offered to take her to the far end of the Keys, to the exotic edge of America, to a flaming sunset over the Caribbean. To where the road ends.

"Oh yes," Bridgett said, her voice full of mock-poetic languor. "Take me to the edge of America. Show me wildlife and wilderness. Throw me into the pounding surf. Crush my bones on the rocks of the shoreline. Make me food for sharks and whales on the edge of America."

Bridgett loved playful exaggeration.

She also liked to invent roles for herself. She had studied psychoanalysis in France. She was very sophisticated. She loved extremes. So we decided to go to the extreme southeastern corner of America and watch the sunset.

On the way to Key West, the sun was hot. Bridgett told me

that she regretted we hadn't met before. "I like you," she said. "We would have had a tropical, Latin passion," she said. "We would have had *un amor brujo*—a burnt love."

Laughing, I promised to throw her into the surf, crush her bones on the rocks, and leave her as food for the sharks.

When we got to Key West, we went to the Duvall Pier to watch the sunset on the edge of America, facing the Tortugas. The daily carnival on the pier was reaching full swing. Tourists were gathering to see the Gypsies and the jugglers, and they, too, were waiting for the sunset on the edge of America. Bridgett pretended to be a Russian aristocrat. I pretended to be a South African just escaped from prison. Once more, I promised to show her magnificent frigatebirds soaring high above a tropical sea.

We sat down on the pier and watched the sun. It hit the horizon and lit up like an Eden-red apple. Then it began to grow squat and flat as it entered the sea.

She watched the Americans on the pier. Bridgett grew satirical. She said that Americans explain things forever. They want a moral and a reason for everything. They are even rationalists about their feelings. Americans are always "coming to terms with their feelings." Americans always need to "clarify things." Americans are always learning to "accept necessary losses." They like to take long baths in front of the mirror, she said laughing, and have kitchens with wide counters and buy lingerie in shopping malls.

People were gathering all around us, sitting on the pier, waiting for the precise moment of sunset.

Bridgett said, "This is the real edge of America—a carnival for tourists with the sun going down."

I had to agree. I looked at the people on the pier, waiting for the sun to set, for the exact instant when it would disappear. They had their cameras ready to validate the moment. I could not imagine them putting down their cameras, or their carnival masks, for any kind of a revolution. I could not imagine them demanding a major change. Looking at them, I felt more like a stranger than a rebel. I did not know where my interest in endangered species would lead me, but I suddenly needed to follow these broken creatures.

"We must go now," I told Bridgett, imitating her mock-poetic tones. "We must find the American wilderness."

"Yes," she laughed. "We do not want to accept necessary losses. We must instead honor our obsessions."

We stood up and left before the sun disappeared beneath the Caribbean. For the first time, I saw endangered species not just as a phenomenon but as a personal truth. They are the lost creatures, and I wanted to side with them. They seemed to offer a much more genuine path into the life of things than anything I had found in mainstream America. For the first time, after the mock-surreal journey with Bridgett to the edge of America, I was willing to choose alienation. Or accept it, as the price for my own view of things. I was willing to walk away from an American carnival.

SIX

GUNS AND PARROTS

I

The mulatto rubbed his crotch and moaned, a low, malicious purr.

"Organ," he said. "Ooooo." He leaned back in the car seat, spread his legs, and rubbed some more.

Oh, my God, I thought, what if he wants sex? Just stay calm.

In the tiny Nissan Sentra, we were almost shoulder to shoulder. Late on a Sunday night, I had arrived in San Juan, Puerto Rico, on my way to the Caribbean National Forest, about 40 miles east of the town. The June night was hot and sticky, the rented car small and stuffy. The plan was to study the endangered Puerto Rican parrot. But my plans for meeting biologists at the forest got screwed up: The headquarters were locked, and the biologists, living high in the tropical forest, behind gates and fences and padlocks, had no telephones. At midnight I was driving down nearly deserted streets in sleeping towns looking for a hotel.

This guy had been walking by an intersection on Highway 3. I pulled over and asked him for directions. My Spanish was worse than his English; we could barely understand each other. Then he leaned in the car window—he was young, his shoulders bare under a tank top—and offered to take me to a *parador*, "the Hotel Caribe."

After a slight pause, and in an impulse of trust and adventure, I let him in the car. It was foreign territory, which always brings out my urge to explore. And besides, I sometimes have this stupid confidence that there's nothing I can't handle.

But now the guy was groaning, I was afraid he wanted sex, and the trip to the rain forest to see Puerto Rican parrots had taken an ugly, malevolent swerve toward the unknown.

I tried to steer the subject away from sex.

"Pee? You want to pee?"

"*Si*, peepee," he answered, and rubbed his crotch some more. He must have thought I was talking about his penis.

We had turned off the highway, down a road leading into the empty coastal plain. We passed through slums, paint peeling on the *gomeria*, corrugated metal roofs rusting. We passed the barricaded houses of snug middle-class families, white stucco behind wrought-iron fences. A few dogs barked at the car, their hollow voices echoing. The darkened houses, with everyone asleep, made me feel like a wanderer after midnight, with no place to stay. Out on the coastal plain, an occasional sad-faced cow flashed in our headlights, sleeping on its feet.

No lights in the distance, no cars, no hotels ahead.

"*Parador?*" I asked, still not comprehending what was happening, still hoping he was taking me to the Hotel Caribe.

"*Si, estray. Hotel Caribe*," he said.

I love the quick turn that takes me toward what I'm always seeking when I head off into nature: fresh and personal vision. I even court it. But it often comes at night, in unexpected guises, when my guard is down on my naked self, enigmatic and laconic as this foreigner.

For the first time, several miles down the coastal plain, I really looked at the stranger in my car. His face was severe, skin taut over cheeks and chin, stubbled with a Latin moustache. The biceps under his muscle shirt were tough and lean. If he tries to rape me, I thought, I'll have to fight.

He lit a Marlboro.

The crazy thing is that whatever was about to happen, I figured

I could talk him out of it. It didn't matter that my Spanish was miserable or that his English gnarled into a nearly incomprehensible accent. The power of language is one of the myths I carry around with me. Words are how I manage my world. My most beautiful weapons.

So I just kept talking to him, almost chatty.

And then he blurted out, "Eside. Heere. Deese eese eete."

I turned right onto a blacktop road, torn up and trashed, weeds growing in the warps and cracks. Looking about, I lost all illusions that we were heading for a hotel. And with that last turn, all my formal reasons for coming to Puerto Rico—studying the Puerto Rican parrot, learning its biology—evaporated in the humid night. What I didn't realize at the time, couldn't realize in my terror, was that he was leading me to a different parrot. They don't give out maps at the airport to the place he took me to, and this detour off the main road would be a blind trip to a very private place.

I screamed at the guy, "This is not the way to a *parador*."

But he screamed back at me, "Estoppe."

I hit the brakes and my mind fell to pieces. The next few minutes were both hyper-clear and incoherent, a series of surreal images like freaks in a strobe light. Strangely, the images came at me slowly, too, the way time slows when it is filtered through the dense medium of fear.

He reached into his pants and pulled out a gun. It shone even in the darkness. He cocked it and held it for an instant poised toward the ceiling of the Nissan.

Hard and metallic, beyond the reach of words, that gun was an incontrovertible thing.

For some reason, it had never occurred to me, fool, that he'd pull a gun. That gun marked with exquisite precision the exact boundary where all sense ends and confusion begins—the truth beneath our contrivances.

And with all the eloquence of pure action, he put it to my head.

A huge rush of paralyzing adrenaline. Me, an idiot in a dumb show. The claustrophobic cabin of the Nissan closed in. The man started yelling something.

Would he shoot?

I slipped into a dark panic. Then suddenly, in the midst of this confusion, I experienced myself in a horrible reversal. We'd switched places all of a sudden, this guy and me. He wasn't the foreigner. This was his country: not just Puerto Rico, but this world of violence. With one quick move, in pointing his gun, he'd taken me to a horrible wilderness on the margins of America.

This night, this side road, this gun at my head were his world.

The stranger was me, coming at myself out of the debris of a wordless terror.

II

I'm sitting in my office in the Pacific Northwest, several months after my trip to Puerto Rico. In my hand, I hold the feather of a Puerto Rican parrot, one I had found later, after my encounter with the gunman, in the refuge of a forest aviary, high in the Sierra de Luquillo. One of its flight feathers, shed and fallen to the floor, it is a small flake of a parrot's life that I picked up and brought back home with me, a talisman of the whole experience of gunman and parrots.

I stare at the feather, trying to make sense of it and the gun at my head, convinced that they're related.

The autumn skies outside my window are a dull gray, not so much ominous as uninspiring, seeping with a melancholy drizzle. I hold the feather between my thumb and forefinger and roll it by its shaft back and forth in my lamplight. When I hold it directly up to the light and look at it straight on, the feather is dull, just a few inches of olive green with black at the base.

But to see this feather, to find its hidden parts, I have to twist it, hold it at oblique angles to the light, look at it this way and that, indirectly, from curious perspectives. I am looking for the sudden insight that makes the moment flare into life, that reveals what the French linguist and philosopher Jacques Derrida calls the "fissure in the allegory" of our apparent truths. When I turn this feather against the light, it flashes into exquisite shades of loveliness, always startling and new in each instant, a piece of molted and molten paradise in my hand.

When I turn the feather just a little, the light glances off its surface with an opalescent luster, brightening with a shimmering, glossy lime, teasing me with a hint of turquoise buried in the green. I turn the feather a bit more, like a kaleidoscope, watching for that wonderful moment when all the colors roll and fall into a prismatic pattern and everything is new. Yellow splashes into the green of the feather, and at just the precise angle, the feather erupts into a fullness of light, a polychromatic feather-flame of burning yellows and bursting greens.

My gray office vanishes, the light takes me completely, and for a brief instant, nothing is but what it seems.

More than one feather dances in the variable sheen of the reflected light. To see the whole feather, I have to look from several angles. What I see depends on how I look. The animals in this book are like this feather: The head-on look, the straightforward empirical glance, is often the most deceptive, especially when we let it come to represent the whole creature. It is true and real, no doubt, but often uninspiring and blind to what it leaves out. It is certainly unable to account for what connects me to animals. I need to roll these creatures around in my mind, study their shifting valences and various meanings, and find just the right angle, where each flashes for me into its full potential and lives most fully in my perceptions.

Now, before I return to the Puerto Rican with the gun at my head, and before I describe my encounters with actual parrots in the wild, I want to turn the parrot around in my mind's eye and see how the light strikes it.

First, the head-on look, the formal and public Puerto Rican parrot. Its scientific name is *Amazona vittata*. The parrot exemplifies the fragility of abundance—the several ways that any species, even species with huge numbers, can be driven to the very edge of extinction. And related to that, it also illustrates the accelerating threats to tropical species and island species throughout the world.

The United States once had four species of parrots. Two are extinct: The parakeet of Puerto Rico died out about 1900, and the last Carolina parakeet died in the Cincinnati Zoo in 1914. A third was extirpated: The thick-billed parrot had disappeared from the southwestern United States, is endangered in Mexico, and is currently being reintroduced into Arizona. Now the Puerto Rican parrot is severely endangered, having dropped to only thirteen birds in 1975, though a disappointingly slow and hard-fought recovery process has brought the population back up to about sixty birds.

The Puerto Rican parrot is not just an image of the past. As far as the Caribbean is concerned, it is in the vanguard of loss—a forerunner of what we can expect on the other islands. Each of the larger islands in the Antilles once bore its own species of parrot. There are nine extant species of Amazona parrot now in the Antilles, all of them endangered; three species have gone extinct. Many other macaws and parakeets have also become extinct. Since Puerto Rico is the most populous and most developed of the West Indies, its history suggests the paths other islands are already following as they break out of their colonial past and grow increasingly dependent on the exploitation of their own resources.

The Puerto Rican parrot is the most endangered of the remaining parrots in the Caribbean, and its loss began with the advent of European colonization. Before the arrival of Columbus and Western culture, Puerto Rican parrots covered the island in huge and raucous flocks, perhaps numbering a million birds. Columbus did not discover Puerto Rico until his second voyage, in 1493, and although by then he had quit describing the flora and fauna, we can infer what Puerto Rico must have been like from his description of parrots on other islands.

For example, on October 21, 1492, he anchored off the Cape del Isleo, a promontory on the north side of Crooked Island. Just over a week since landfall in the New World, Columbus examined the island and lost himself in wonder. His vocabulary beggared by what he saw, Columbus felt unable to describe the pristine beauty and pure enchantment of the place, which he compared to "Andalusia in April":

The singing of little birds is such that it seems that a man could never wish to leave this place; the flocks of parrots darken the sun, and there are large and small birds of so many different kinds and so unlike ours, that it is a marvel. There are, moreover, trees of a thousand types, all with their various fruits and all scented, so that it is a wonder. I am the saddest man in the world because I do not recognize them.

Note again, as with manatees, the impulse of Columbus to liken what he finds to the Old World, to try to orient himself in terms of the language of the old—in terms of what is familiar. In the New World, Columbus found huge flocks of parrots and a certain inability to "recognize" what he was seeing.

On his first day in San Salvador, Columbus described the natives carrying parrots as pets as they swam out to his boats, offering to give the birds to the Spaniards. Having parrots as pets was apparently a custom throughout the islands, including Puerto Rico. The natives also used the feathers for ornamentation, fletched their arrows with the green plumes, and even ate some of the parrots. According to one source, the natives supposed that parrot meat enhanced a man's sexual potency.

On Puerto Rico, too, parrots once flew in staggering flocks. From the sketchy records of explorers, it seems that parrots lived throughout the island, from the rain forest in the east to the dry karst forest in the west. Puerto Rican parrots also occurred on three of the four tiny offshore islands: Vieques, Culebra, and Mona. The flocks were described by explorers as "dense," "great in the interior," and "abundant," and they apparently were capable of a deafening clamor. Through extrapolation, biologists have arrived at the estimate of over a million Puerto Rican parrots before Western colonists settled on the island.

The original Tainos Indians probably had little impact on the parrot populations. It remained for the European colonists to bring mass destruction, a by-product of their civilization.

The main cause is the familiar one for the animals that have become endangered at the hands of humans: habitat loss.

In Puerto Rico, that meant deforestation. Although originally the island was completely covered with forest, by 1828 about a third of the forest had been cleared. Most of the loss happened in the next century, spurred by a burgeoning population. Between 1800 and 1850 the island's population tripled, and then it doubled again in the next fifty years. By 1900, the population was about a million people. In 1899, only about 8 square miles of undisturbed forest remained.

The Spanish crown cultivated most of its land and encouraged deforestation for agriculture. In the late nineteenth century over three-fourths of the island, even low-production land, was converted into farms for coffee and sugar, and many of the people doing the farming were poor, not owning the lands they worked.

By 1912, not more than 5000 acres of virgin timber remained on the island—less than 1 percent of the original forest.

During this century, the parrots clung to various refugia in the rugged mountains, mostly degraded forests that were too small and too disturbed to succeed as sanctuaries. These remnant patches, shreds of a tattered paradise, vanished one after another. In Guajataca, a limestone karst on the western end of the island, the parrots disappeared by the 1920s. In Rio Abajo, the last western holdout, the parrots disappeared by the 1930s. In Sierra de Cayey in the southeast, the parrots were last reported in the mid-1930s.

Finally, parrots could be found only in the Caribbean National Forest of the Sierra de Luquillo. Their current range is about 1600 hectares, or 3953 acres—about 0.2 percent of their original range.

Even during this century, it has been very hard to keep up with the bird's steady loss. What happens to animals is one thing, but how we comprehend what happens is another. Even as biologists were beginning to realize how serious the loss was, how complete the devastation, we have had trouble realizing what it meant.

The light off the feather is shining brighter now for me.

Nor was it just the rape of the forests that destroyed the parrots. Other violent losses occurred too.

The Puerto Rican parrot was once a common pet on the island, and parrots remain an extremely popular pet in Latin cultures, where they are often allowed to fly free around the house. There is also a huge, wealthy passion for rare parrots in the United States and

Europe. From the 1930s to the 1950s, local parrot traders stole chicks from at least four of the five nesting areas in the Sierra de Luquillo. Robbery was rampant in some areas: Ten known parrot harvesters stole six to twelve nestlings per man per year. Nest robbing was a major cause of the decline of the parrots, and by the 1960s people were stealing parrots from nearly every nest in the mountains.

The trade in parrots, legal and illegal, is big business. Endangered species are supposed to be protected by law, but the demand for pets, charming as they are, puts enormous monetary incentives before the poor. A fairly common parrot can go for $500 to $2000 in a pet shop, but an endangered species such as the green and purple imperial parrot on the island of Dominica in the West Indies may fetch tens of thousands of dollars. One researcher reported hearing a price of $30,000 for a parrot. A U.S. Fish and Wildlife Service investigation documented illegal trade in parrots: As many as 26,000 parrots per year are smuggled across the United States–Mexico border near Brownsville, Texas, alone. The thick-billed parrot was extirpated in the United States, almost surely by being shot, is currently stressed by the pet trade in Mexico, where it is now endangered. Ironically, the stock for the current reintroduction project in Arizona is provided from confiscations of parrots illegally smuggled into the United States for the pet trade.

No parrots are now stolen in Puerto Rico, since biologists watch the last four active nests with constant vigilance. But in stealing the chicks, robbers sometimes took the lazy way, cutting down the nest tree, and thereby also selectively destroying the nest sites for future breeding.

Other human pressures on the parrots had the effect of destroying the birds' nesting habitat. Their favorite nest tree is the palo colorado, which was a major source of charcoal for indigent locals. As recently as 1949, 29 percent of the farmers on the island continued to rely on charcoal for cooking, and 43 percent on wood as fuel. One study between 1934 and 1937 estimated that an average farm would use 111 sacks of charcoal per year, and at that time there were over 55,000 farms on the island. The precise impact on the parrots of cutting the palo colorado trees is hard to establish, but it seems clearly to have affected the quality of the forest and the availability of nest sites for the birds.

The locals also cut down trees for honey, which is common in the Sierra de Luquillo. The honeybees use the same kinds of cavities the parrots nest in, and cutting down honey trees probably also reduced the number of suitable nest sites for the parrots.

Finally, shooting. According to a Ph.D. dissertation by Frank Wadsworth, who later became forest supervisor for the Institute of Tropical Studies, the Puerto Rican parrot used to fly out of its refuges in the mountains into the surrounding farms to feed. Considered harmful because they eat young corn, plantains, oranges, and other fruits, parrots were driven from the fields. As late as 1903, farmers were still shooting them.

The Carolina parakeet in the southeastern United States was also fond of farmers' fruits, and no less a personage than John James Audubon called their depredations "outrages":

> The gun is kept busy, with eight, ten, or even twenty being killed at each discharge. As if conscious of the deaths of their companions, the living birds sweep over the bodies, screaming loudly as ever. . . . I have seen several hundred destroyed in this manner in a few hours.

Such ravages sent the Carolina parakeet into extinction.

And here, as you might have guessed by now, is exactly where the feather flashes brightest, grows luminous in my hand. Here is where I can begin to see and suffer myself in the parrot. Robbery. Guns and parrots. Guns and me.

Carolina parakeet, thick-billed parrot, Puerto Rican parrot: It is not hyperbole, and it is more than metaphor, to say that we have put a gun to the head of the parrots in North America.

Before any of us gets too complacent, imagining that matters are now under control or that this is a problem confined to a small island like Puerto Rico, let's shift the light on this subject a little and see what else appears.

The destruction of the tropics throughout the world is a problem receiving considerable press. But the problem is urgent and growing, and the destruction is of such a magnitude that it seems almost beyond comprehension, much less easy solution.

We don't even know how many species of plants and animals there are in the world, and the number keeps getting revised upward. According to Norman Myers in *The Sinking Ark*, in the 1960s the number of species on earth was estimated at around 3 million. By the early 1970s, a new figure was proposed: 10 million. It's a crude guess, but as Myers writes, "It helped scientists recognize the scale of the challenge they were facing: how to preserve the panoply of the earth's life forms." Of these, about 160,000 species have been identified and named.

The actual number of species may be even larger and more difficult to fathom than these estimates, however. According to Edmund Wilson in *Biophilia*, an entomologist has estimated that there may be 30 million species of insects in the world, most of them in tropical forests.

Tropical forests, of the sort lost in Puerto Rico, have the greatest abundance and diversity of species in the world. Yet we continue to clear them at a prodigious rate According to Norman Myers again, using what he calls "rough and ready figures," we could be losing as much as 50 hectares (about 123 acres) of tropical forestlands *per minute*. Estimates on the rate of loss of tropical forests vary considerably, but they are all very large, and accelerating.

All this habitat destruction translates into lost species. As Myers writes, again using estimates, "Let us suppose that, as a consequence of this man-handling of natural environments, the final one-quarter of this century witnesses the elimination of 1 million species—a far from unlikely prospect. This would work out, during the course of 25 years, at an average extinction rate of 40,000 species per year, or rather over 100 species per day."

North Americans are probably not innocent of the deforestation in the tropical regions of Central and South America. The impacts of North American companies and North American demand are not clear—this topic would make an excellent book-length study. Some people, however, claim that the impacts are large. As Jack Connors has written in a recent article on an "ecological tragedy in the making," "The enormous American consumption rate of coffee, bananas, sugar, and beef makes them major exports [for the economies in the region], and the production of each has required the clearing of vast tracts of forest."

The Puerto Rican parrot also represents the problems facing endemic species on islands. *Endemics* are species that have evolved nowhere else, and islands such as Puerto Rico and the Antilles feature large numbers of endemics. The isolation promotes speciation. In the geography of extinction, island fauna are the most vulnerable. Of all the bird extinctions in the last few centuries, fully 90 percent have come to island species. According to Paul Opler, a biologist with the U.S. Fish and Wildlife Service, of the 518 extinctions (species and subspecies) in the United States up to 1977, 351 (67 percent) had been to island dwellers. Hawaii accounts for most of them, but twelve species and subspecies have also been lost from California's Channel Islands.

Hawaii is a classic example of island extinctions: The plants and animals did not develop defenses, and they have been quickly destroyed by humans and introduced species such as pigs, dogs, cats, rats, cattle, and mongooses. Since the arrival of Captain Cook in 1778, at least ninety species of indigenous birds have occurred on the islands, plus another hundred or more migrants, or vagrants. At least twenty-three of the passerine, or perching, birds (40 percent) and both of the Hawaiian rails have gone extinct in the last century; of the remaining thirty-four passerines, twenty-three are classified as endangered. This means that 80 percent of the Hawaiian passerines discovered in the last 200 years are now extinct or endangered.

The rest of the birds in Hawaii, though not yet classified as endangered, face very real threats to their existence. U.S. Fish and Wildlife Service biologists J. Michael Scott and John L. Sincock undertook a major study of Hawaiian birds. They have concluded that Western culture bears primary blame: "Human-caused extinction of Hawaii's unique birds began with the Polynesians, but it was accelerated vastly after the arrival of western man."

Ironically, we can expect rates of extinction to rise in our national parks—the places we set aside for the creatures' protection—

by virtue of their isolation and distance from new animal colonists. National parks and wildlife refuges are artificial islands—natural systems that are surrounded, for all practical purposes, by seas of transformed countryside. The future of many animals in national parks is also their peril.

Western culture is exporting this catastrophic drama around the world, with devastating effects, wherever it comes into contact with vast stretches of unexploited resources. The terms change in this global drama—if it's not pet stealing, or charcoal, or shooting, or the introduction of exotics, it *is* something else. The particular causes change from place to place, though habitat destruction seems to be a constant. The characters change too. But the roles and the plot are always the same.

III

The Puerto Rican started screaming at me to get out of the car, and I didn't know this car at all, and I was bumbling and blundering around, grabbing for the door handle and the emergency brake and the lock on the door. He didn't want me reaching for anything, so he shoved the gun hard against my head as we both grew more frantic. The last thing I wanted was for him to panic and start shooting.

Both of us were nearly out of control.

When I got the door open, I almost fell out of the car onto all fours on the blacktop. Finally, I was standing on the ratty pavement.

He bolted out of his side of the car, still yelling, something about money. I took out my wallet and almost threw $200 at him.

He came around the car, jammed the gun into my gut, took the money, and grabbed my wallet out of my hand. Then he put the gun in my face and yelled at me to run. He was so close that he spit in my face as he shouted, and I remember his teeth.

I just stared at him blankly.

Now what? Was this the moment to fight for my life, since he might shoot while I ran?

Then he shoved me with the gun. "Run," he yelled again.

I ran. Or loped really, about 20 feet, looking back over my shoulder at him in his painter's cap. He yelled at me again to run.

And now, finally, I knew what he was going to do.

I stopped running and looked back at him, standing beside the car, the driver's door still open. I put my hands up, pleadingly, and said please, please, don't do this. As if anything I said would do any good at that point. But I begged anyway.

And then he did just what I had realized he was going to do He jumped into the car. The brake lights flashed red at me as he put the Nissan in gear. The engine revved, the wheels popped on the pavement, and he took off.

He took everything I had brought with me on the trip—clothes, I.D., notes and journals, part of a chapter for this book, optical equipment. None of what I had lost was quite clear to me at the moment; I felt just a generalized sense of loss. But it would all come to me, slowly, item by item over the next few hours.

Nor did I feel lucky that he hadn't shot me.

I was alone in the dark on some deserted back road. No possessions, no money, no I.D.—a nameless foreigner in poor territory, nearly a nameless corpse in a ditch. And the worst of the darkness was not the night but the confusion and fear inside me.

I stood there for a few moments, silent. Then I screamed like mad, jumping up and down: "You idiot, you goddam fool!"

The night got oppressively hot, and I got a second rush of panic. Abandoned.

I wanted desperately to do something, get out of there, start immediately to put everything back together, get some control. I longed to be in that mythic world of Someplace Else.

And I did what I have often done in a crisis when I've been really scared. I ran.

I ran all the way back to the highway, about 10 tumultuous miles, nonstop. I could feel at least some power in the muscles of my legs, some strength from being in shape. One minute, while running, I felt like an utter wimp, and the next minute I was filled with murder, ready to kill that guy with absolutely no regrets. Bash his face with a baseball bat.

At the highway, I waved at cars to pick me up. They drove right on past, like I was some monster out of the jungle, soaked with sweat and dirty. No one would pick me up, afraid no doubt that I would do to them what the Puerto Rican had just done to me.

I walked aimlessly along the highway, feeling as much like the outsider as I have ever felt, completely cut off from everything.

Every time a car came by, I waved my arms, even jumped up and down, hollering and pleading for it to stop, though I knew the driver couldn't hear me. After about an hour, a police car came along, its light bar flashing bright blue. It passed; then red brake lights lit up, the car slowed, and it pulled to the shoulder. Before it stopped, I was sprinting to catch up to it. I sank with relief into the backseat and slammed the door closed.

The two police took me to a local headquarters in a small town, where I immediately became the main attraction. It was 3 a.m., roosters crowed in the yard outside, and a couple of dogs barked. All the police in the station, and a couple of locals who for some reason were there, talking, gathered round and stared at me.

I didn't know where I was or what was going on. The police seemed to mean well, but they didn't even ask for a description of the robber and I didn't know their language. I was sure I'd never see any of my stuff again. I got this vivid picture of myself, an American in the Third World, reduced for these people to the spectacle of another stupid gringo who'd been dumb enough to get robbed.

IV

By the time I got to the Caribbean National Forest and met the people who would introduce me to the actual parrots, I had had enough of guns, of strangers, of reality altogether. Those feelings of openness and trust, of dialogue with the manatee, now meant nothing. Trusting too much had gotten me into the mess with the gunman. All I wanted was a sense of sanctuary—to escape into the forest, spend some time with the parrot, and try to forget what had happened.

When I got to the forest, I hooked up with Wayne Arendt, a biologist on the recovery team. He was full of Latin energy, the result of eight years on the recovery project for Puerto Rican parrots and a stint before that with the Peace Corps in Hispaniola, where he'd married a woman from the Dominican Republic. Though Wayne was not the director of the project, he was working on his Ph.D. and spent long hours in the forest with the birds. He took me under his wing and helped me see the parrots.

Established in 1968, the recovery project for the parrots has been an intense effort, in many ways a model of conservation biology, especially in the tropics. James Wiley was the director of the project when I visited. It is a cooperative effort among the Puerto Rican Department of Natural Resources, the U.S. Fish and Wildlife Service, and the U.S. Forest Service. From a low of thirteen birds in the wild, the biologists had slowly built the population back up to twenty-eight birds at the time. In addition, they had established a captive flock in an aviary high in the forest, and it had slowly grown to about thirty birds. The Puerto Rican parrot has proved hard to work with in almost every stage of its recovery, and nothing, including captive breeding, came easily. It was a constant struggle, a fight for each new addition to the flock.

When Wayne learned what had happened to me on the way to the forest, I became a living embodiment for him of all he hated about Puerto Rico. His black hair hung straight along the sides of his balding head and shook when he talked. He was out front with his feelings, and I sparked all his impulses to get out of Puerto Rico. Partly, there were morale problems on the project, the result largely of competition among the agencies overseeing the recovery. There was also some burnout after so many years of the constant pressures of monitoring and helping a bird as intractable as the Puerto Rican parrot. But the big reason that he disliked Puerto Rico was the crime.

We both wanted Out.

For Wayne, that meant out of Puerto Rico; for me, into the forest. It gave us a bond, and he went out of his way to help me. Since the parrots are easily disturbed, outsiders are not normally allowed into the forest with researchers. But Wayne had work to do in the general vicinity of a nest, and he managed to finagle permission for me to join him.

We pitched into the rain forest down a steep trail in a gray dawn. From the outside, in a panoramic view, the forest looked like a strangely beautiful fever—the bulging, round tops of the trees like green swellings on the convulsions of the mountain's hills. But when you enter the forest, its character is nearly tangible, weighty.

Spongy clouds had closed upon the mountain—we were at about 4000 feet—and the air had a gelatinous feel to it, drippy as egg whites. Tree roots swarmed promiscuously across the trail, snarling around jagged basaltic rocks, and both roots and rocks were slippery and treacherous from overnight rains. Since there is little topsoil in the mountains, the roots tangle across the surface. When the trail broke free of the roots and rocks, it was mud. Ankle-deep and the color of cayenne, it splashed up several times into my face like an affront.

Overhead was the mass of the canopy, stitched together with vines and bromeliads.

This tropical rain forest was fairly benign—no dangerous spiders, no deadly snakes. But it did have a recalcitrance to it, a toughness not easily overcome. Its physical challenge was felt in the strenuous labor of hiking through it. Clinging to trees and vines for balance, I worked up and down several steep ridges. At one point we crossed a magnificent cascade of boulders, huge rocks that formed a streambed for a small trickle of water that swelled into a river of waterfalls in the torrents of rain in the forest. Worn by the water, the stones looked like an arrested avalanche. We crabbed our way across them on all fours, feet down, hands back, butts dragging. Following the stones in a couple of streambeds—ankle-twisters covered with squashy mosses—we scaled a nearly vertical hill to the ridge above the nest.

My quadriceps ached from the work, my body was drenched with sweat, and I felt great. We slumped our packs on a trail that ran north and south on the saddle of the ridge.

Wayne was grumbling about the hike in, about being there in general. For Wayne, life was, well, full of problems. He loved to complain about them, and decisions weren't easy for him. A com-plicated soul, he was so open about his problems that he was charm-ing. The hikes on a tough trail were a pain, and the work today was a distraction from research on the Ph.D. he needed to finish.

"Once I loved the challenges here," he said, "but not now."

While he got the radiotelemetry gear together for an experiment, he heard the parrots fly in, squawking with the same hoarse screeches that have probably deafened you in pet shops. Through a couple of small breaks in the canopy of the trees, against the gray swatches of sky, I caught a fleeting look at two parrots, just chunky dark shapes whose hardworking short wings pounded toward us with choppy strokes.

That glimpse whet my desire. In the quiet after the birds' honk-ing, the forest got heavy again, the air weighted and wet. Wayne couldn't resist teasing me: "You've seen parrots. But now I suppose you'll even want to see green on 'em."

More squawks.

Wayne explained that he knew they had landed by their calls. In flight, he said, they herald their approach by a throaty honk, honk, honk. When they land in a tree, the call modulates into a more nasal wank, wank, wank. That's what they had just given, and the sound had a jagged inflection to it that put a bite into the muffled heaviness of the morning.

Wayne got to work and told me I could look for the parrots. He warned me to be quiet, to stay close to the path, and not to bother them.

Out of a grotto of old trees, tangled into a mass of vines and twisting branches, the parrots squawked again. Since they were so rare and so easily disturbed, I tried for stealth. This was the moment I had come to Puerto Rico for, and my heart was pounding hard and fast. The almost tangible green air of the forest was like a velvet curtain in front of my eyes. The parrots were no more than 20 yards away.

They squawked again.

I knew which tree they were in and tiptoed toward them on the path, ducking branches, searching for precarious footholds in the roots. I scanned the tree intently, but couldn't see them through the profusion of the trees. By now, the thick air and heavy silence of the rain forest had grown oppressive to me. I knew something was there, right before my eyes, yet I couldn't see it. It was as if I were part of an audience, waiting for a play to begin, but no one came on stage—I was waiting and waiting and nothing was happening. Expectation grew into a burden.

I took another quiet step around the base of a tree.

With an explosion of screaming, the parrots tore right into the fabric of the moment, ripping open the silence of the forest. There I was, trying to sneak up on them to avoid disturbing them, wanting only the peaceful experience of seeing them in the heart of their refuge. But they had begun to scream like a burglar alarm, an unbelievably loud cacophony only 30 feet away, and just trying to see them seemed like a strangely dangerous business.

Their wailing gave me an intense rush of self-consciousness. I had the sense of being there, knowing something was happening right in front of my eyes, but being blind to it. Even when I'm surrounded by nature, I can feel that something is concealed from me, hidden just beyond my eyes, and I'm flailing against the thick curtains of a stage, looking for the opening that will let me through.

In an awful wave of desperation, I realized I *had* to see those parrots. There was some epiphany in the trees, some secret, and I couldn't stand the idea of being there and missing it. I just kept my eyes on the tree they were in, hoping to see something.

As I scanned the tree, I saw one of the parrots hop from one branch to another. Its back to me, it took a slow parrot-step sideways along a branch and reeled around. No longer screeching, it sat there staring down at me, silent and steady as a gargoyle.

This wasn't much of a view really. The light was bad, the air foggy, the parrot not much more than a dark shape up in the tree. Colorless, mostly, though I knew its plumage was, in good light, a brilliant and varicolored green. About 12 inches long, with a short head and no neck, thickset. I could just make out the white eye ring on its face.

Still, I live for moments such as these, that exact instant when it seems as though a veil is being lifted. Seeing a new animal in a forest for the first time is for me as if a secret is being revealed. If life is a journey through unfamiliar territory—through dangerous cities and dense forests—every once in a while it offers beautiful views.

Truman Capote recounts a dream about a parrot in his book *In Cold Blood*; the dream comes to Perry Smith, one of the two murderers. In it, a parrot lifts Smith out of the violence and frustrations and agonized squalor of his life into a world of blue oceans and long tables laden with oysters and fruit, where Smith can finally say, "I know where I am." The parrot first flew into Smith's dreams when he was a young boy living in an orphanage run by nuns—shrouded disciplinarians who whipped him for wetting his bed. The parrot lifted him into paradise. Later, Smith dreamed that the same parrot carried him out of prison as he was awaiting execution for his murders.

Perry Smith's dream, and the hazy parrot I saw, were examples in extremis of what I'm often looking for in nature: the perfect illusion The ideal Out.

I did not get to watch the parrot for long, maybe two minutes. It took off, and another with it that I hadn't seen. They left with their staccato departure call, wak-wak-wak, which drifted through the slow air like falling feathers. I could hardly wait to get back and tell Wayne what I'd seen.

V

Filled with a sweet expansiveness, I walked back to the saddle, where Wayne was working with the radiotelemetry experiment. When I got there, Wayne was upset. The experiment was not going right, but mostly he did not want to be in the forest that day. He had reading to do for his dissertation, and this experiment, he said, was grunt work.

I sat down on a rotten log, happy to be there.

Wayne pointed to a rainstorm heading our way. We could hear

it coming across the ridges, getting louder, a deluge that beat onto the canopy of the forest with the gathering roar of an approaching train. Wayne hustled us to the protection of a nearby blind, one he had built years earlier to watch a nest that was not being used this year. In a matter of minutes, the rain was pounding the blind like bricks.

The Puerto Rican parrot is not only a symbol or a creature of somebody's dreams. Each one comes to us out of a tangible world of painful particularity, with its own precarious history. The list of the biologists who have worked to save the Puerto Rican parrot reads like a *Who's Who* in conservation biology: Cameron Kepler, famous also for his work on Hawaiian birds; Noel Snyder, whom I had met when researching the California condor; and James Wiley, who was the project director when I was in Puerto Rico. The work of these biologists honors each creature as a living being, bears witness to its story and distresses, and champions a new contour to its future.

Though the Caribbean National Forest provides a refuge for the last of the Puerto Rican parrots, it may not be the best habitat for the birds. Even after the nest robbing and the harvesting stopped in the 1960s, the parrots still faced severe and, at the time, unidentified problems in this forest. These turned out to be, in one way or another, "natural" problems, not the result of direct human persecutions. But they show just how fragile a population of animals becomes when driven to the edge of extinction. In fact, without the monitoring and care of the biologists, the parrots could vanish tomorrow.

One of the big problems the parrots faced was the rain. Out the portal of the blind, I could see a gnarled old palo colorado tree, its massive trunk mottled in ochers and oranges, its bark flaking. With its twisting profusion of branches, the palo colorado offers deep cavities for parrots when some of its branches die and fall off. Not much grew under the tent of this tree's domed top, and the ground beneath was littered thick with its browned leaves.

The tree outside the blind was testimony to the ingenuity of the biologists, both in figuring out why the parrot was having trouble and in doing something about it. Looking anything but natural, it had been turned into a kind of federally sponsored housing project for parrots. It reminded me of Winnie-the-Pooh's house built into a big tree. The actual hole for the nest was about 15 feet off the ground, in a branch on the right of the tree. Above the hole was a metal awning, painted black. Parrots like to nest deep in the cavity, 6 to 8 feet below the hole. ("They must feel safer," Wayne said.) A door had been built into the tree at the base of the cavity. There were also rectangular sheets of black metal in several places, like back doors and windows. And the whole business had been caulked and painted to match the tree. The nest was fully insulated from the rain.

One study in the 1950s found that only about 20 percent of the nests that got to the egg stage actually fledged. When Noel Synder came to work on the Puerto Rican parrots in 1972, he realized that this was extremely low nest production for a hole-nesting bird, and he viewed the nesting biology of the species as "one of the critical weak links."

In an effort that is hard to imagine, given the rugged density of the rain forest, Noel and other biologists scoured the 28,000 acres for nest holes, examining over 1000 nest trees in the nearly impenetrable slopes of the mountains. He found few suitable for the parrots.

Noel and others spent thousands of hours watching the nests. The problems began to take shape in their minds. They watched females emerge from cavities, after incubating, with muddy breasts and wet feathers. Others used cavities that dropped through rotten wood clear to the ground, where the parrots laid their eggs. One pair even tried to nest in a tree that was hollow to the ground and open at the base. Sometimes, parrots were seen fighting over a nest site—which neither pair wound up using.

Noel decided that there must be a shortage of dry nest sites. As a result, he started the active intervention that has led to the renovated nests I was looking at through the blind.

Of the four nest sites still used by the parrots in the Caribbean National Forest, only three were active the year I was there. Each of the active nests is monitored every day, from before dawn to after dusk, and information is gathered on the biology of the birds—growth rates, fledging times, the successful production of the nest, that sort of thing.

As we sat in the blind, Wayne told me a story about James Wiley, the project director, and three chicks as they were fledging. The chicks emerged from the nest. This is one of the most critical and vulnerable times for the parrots, so they keep extremely silent for fear of predators. Because rain had seeped into their nest cavity, the chicks' wings were brown, not green, caked with wet wood. With the parent watching discreetly from a nearby branch, two chicks, apparently able to fly, fluttered on wobbly wings to another tree. The parent followed.

But the third chick toppled to the ground. A volunteer observer rushed out, grabbed the parrot, hiked out of the forest, and took it to the aviary. There, James Wiley performed emergency surgery. Clipping off the flight feathers, he collected new ones from the floor of the cages of other captive parrots and glued them to the chick's wings with nontoxic epoxy. Back at the nest, the parrot later fledged and grew new flight feathers.

Wayne has found that chicks in the nest can be infested with the larvae of warble flies, grubs that bore into the birds' legs and head. These chicks are pulled from the nest and treated in the aviary. Eggs are also pulled when necessary, and dummy plaster eggs are even placed in the nest to keep the parents in the practice of incubating.

A second major breakthrough in the recovery of the parrots pertained to a natural predator, the pearly-eyed thrasher. A nondescript bird with a call like a crow or jay, the thrasher is now abundant in the forest. Wayne's Ph.D thesis is on this species. At the turn of the century, the pearly-eyed thrasher was a rare bird in the Sierra de Luquillo, but according to Wayne, it is an example of an aggressive generalist species that flourishes when a habitat is traduced—much like house sparrows, starlings, and mute swans in the United States. Good competitors, this type of bird puts enormous pressure on more sensitive species, driving them out. On an island, birds like the thrashers are especially tough on endemics.

"After the forests are raped," Wayne said, "pearly-eyed thrashers can prosper in fields and yards and parking lots as well as forests."

Noel Snyder noticed that parrots and thrashers begin searching for nest sites at about the same time, between January and March, and that they compete for similar tree cavities. Also, thrashers are often still looking for nests when the parrots have chicks. If the parent of the chicks is at home, the thrasher moves on, leaving the babies unmolested. But parrots are not constantly on the nest, and a thrasher that finds an unattended nest will eat the eggs or chicks. In 1975, Noel found that every parrot nest was harassed by thrashers. Over the years, five nests failed because a thrasher got eggs or chicks.

As he did when he took eggs from condor nests, as he did when he modified parrot nests, Noel intervened. He solved the problem by providing nests for the thrashers. He placed shallower nest boxes near deep, dark parrot nests so that the nesting thrashers became a "guard pair"—keeping other thrashers away but not threatening the parrots' nests, since they had their own. Thrashers have not been a significant problem for parrots since.

James Wiley feels that most of the major problems facing the last remaining parrots have been solved. The current goals are to establish a second wild flock in another location, relieving some of the pressure associated with having only one group of birds. The parrots remain extremely vulnerable to natural catastrophes—hawks preying on chicks; a warble fly infestation; or a hurricane, which in Puerto Rico can be devastating. The hurricane San Cipriano, of September 26, 1932, smashed into the north and east of the island, killing 225 people; according to an eyewitness, "all the trees were stripped of their leaves." We can only guess at the death toll for the parrots, and all the researchers worry about what will happen when another huge hurricane strikes.

So the parrots are a fragile triumph, and a guarded optimism pervades the project. The big question: Will the agonizingly slow, one-parrot-at-a-time recovery continue indefinitely, or will the population suddenly boom? All the harvesting of forests and baby parrots left a largely geriatric population for biologists to work with, but larger numbers of young parrots are coming into maturity, which happens at about 4 or 5 years of age, and the hope is that the flock will hit a critical mass and its population will take off.

In the blind, Wayne and I waited out the rainstorm, which lasted for about half an hour. After it passed, the forest began to steam in the midday heat like a closed aquarium. Wayne told me to stay up on the ridge and wait through the afternoon because, with a little luck, I might catch sight of the parrots again when they returned to their roost for the evening. He took off to work on his experiment some more.

I felt an enormous affection for Wayne, with his complicated soul and his grumbling, dogged devotion to the parrots. He had done a lot for the birds when they were in trouble, and he had done a lot for me when I was in trouble. These feelings extended to Noel, also, and to James Wiley. These biologists had saved the parrot with insights, commitment, and action. There are a huge number of people, throughout the United States, working hard and anonymously on behalf of endangered animals—I've met many of them, and they deserve much more credit and recognition than they get.

This business of helping was not some abstract principle to me as I stood in the forest. The clothes I was wearing, Wayne had lent me. Wayne had supplied the binoculars I was using. Wayne and his wife, Angela, had put me up and fed me. I was in the forest because of Wayne's help.

I walked back to where I had seen the parrots in the morning and settled in at the base of a tree on an overstuffed sofa of club moss. The sky cleared; the air heated up. I dug my heels into the ground and smelled the richness of the humus. Wayne had said that the parrots would probably return to the area in the late afternoon, to roost. I waited for them.

VI

I never got any of my stuff back.

After taking my story, the police wanted to get rid of me, so they more or less abandoned me at a hotel at four in the morning.

No money. No credit cards. No I.D. No way to check in. They took me to the lobby. I went to the bathroom, and when I came out, they had vanished.

But the night clerk, a round-faced man named Gilberto Rodriguez, gave me credit. He found me a room, and room service too. In the morning, he waited around for me to get up and then drove me into San Juan.

In town, the supervisor at the car rental agency, a robust woman named Nina Cramer, went far out of her way for me. She got me another car and let me tie up her phones making long-distance calls all day. Then she stood reference for me at the bank, where, after a lot of talking, I managed to get a few hundred dollars. She did the same thing at the airline, as well, where she helped me convince the clerk to let me charge a ticket on a credit number, without a card or I.D.

After I got my ticket, I went to the phones to tell my relatives I would be all right. The phones in the San Juan airport are not in private booths but out in the open, forming a sort of island in the middle of the huge, verandalike lobby. The place was swirling with people coming and going. Vacationers in bright tropic shirts, open at the collars. T-shirts splashed with the painted logos of one or another of the Virgin Islands. Latin men in their white and lacy shirts, untucked at the bottom. Families with rowdy kids. Occasionally, conspicuously, nuns in black and white habits. It was two in the afternoon, and I was filthy, my pants and shirt reeking from an awful night and day.

On the phone, I told my mother about the police station and about how helpless I had been. Telling her what people had done for me, I started to break down for the first time. Amid all those people milling around me, transient but purposeful, on their way to specific places, and amid people waiting for the phones, I lost control. I started to sob on the phone.

I didn't cry for the chaos of that night or because of the gunman. The tears came when I thought of how dependent I had been on other people. The tears came when I let myself feel gratitude for those who had helped me when I could do nothing for myself.

VII

I have had to struggle repeatedly with the underlying question of this book: Is everything fiction? A role? Is every image of nature, and of ourselves, a construction of our minds, a product of our cultural and personal histories? Is it possible to get beyond ourselves to some genuine perception of nature, to see a real animal, not just an animal we've partially invented in our minds, like Columbus with his sirens or us with our abstract and distant creatures? Is even the image I had of myself after the robbery—the outsider with the storm raging within—just one more fiction, borrowed from modern literature? Is the notion of endangered species as outsiders simply a reflection of my own self-image?

Perhaps the best we can ask for is to live happily inside our own illusions, knowing them better but never escaping them. I'll let you live in your dream; you let me live in mine.

Beneath these questions are some of the major philosophical issues of our century. I have argued so far that biology may claim to offer realistic descriptions of animals but that that realism is an illusion of its own. At best, the biological description is one vision among many, partial and not complete. We are anesthetized by it, though, forgetting that it is not the whole story, forgetting even that it has in itself cut us off from the creatures we try to know. It is a kind of epistemological imperialism.

But is it possible to have *any* relationship with animals, with other people, with our own selves that is immediate, real, unmediated by image or language or role? Is this not, finally, the great task: to determine how we will live, and to establish a genuine relationship with those with whom we share the world. The great French anthropologist Claude Levi-Strauss describes in *Tristes Tropiques* the confusions and agonies of the attempt. Traveling to South America, he wrote that he "had wanted to reach the extreme limits of the savage." This of course is the point: to journey to the heart of things and come to the virgin land. But Levi-Strauss says that the savages "retained their strangeness," and he was "incapable of even grasping what it consisted of."

Then he writes: "But if the inhabitants were mute, perhaps the earth itself could speak to me . . . , perhaps it would answer my prayer and let me into the secret of its virginity. Where exactly does that virginity lie, behind the confusion of appearances which are all and yet nothing." This is the paradox that four centuries of rationalism have led us to: For all our efforts to know the world, we remain, like Columbus, wanderers in a territory we cannot comprehend.

One lesson of twentieth-century linguistics, which has been closely allied in its efforts with anthropology, is that we do not simply perceive the world. We interpret it. Humans live locked in a world of language, excluded from the natural world by the language that gives us access to it. In *Structuralism and Semiotics*, Terence Hawkes gives an excellent summary of the development of this line of thinking in modern linguistics. Largely through the work of a Swiss linguist named Ferdinand de Saussure, we have come to realize that language is a self-sufficient "system of signs." The relationship between the concept of a tree and the sound image made by the word "tree" is a linguistic sign. These signs combine to make up a language, and they are arbitrary. As Hawkes writes:

> The overall characteristic of this relationship . . . is arbitrary. There exists no necessary fitness in the link between the sound-image, or signifier "tree" . . . and the actual physical tree growing in the earth. The word "tree," in short, has no "natural" or "tree-like" qualities, and there is no appeal open to a "reality" beyond the structure of the language in order to underwrite it.

In the end, language acts as a conservative force in the human apprehension of the world, and constitutes its own reality.

Such a view of language implies that our lives as individuals are social to the roots, that the world as we know it has been structured by an inherited language. The American linguist Edward Sapir writes in a famous passage:

> Language is a guide to "social reality." . . . Human beings do not live in the objective world alone, nor alone in the world of social

activity as ordinarily understood. . . . It is quite an illusion to imagine that one adjusts to reality essentially without the use of language and that language is merely an incidental means of solving specific problems of communication or reflection. The fact of the matter is that the "real world" is to a large extent built up on the language habits of the group. No two languages are ever sufficiently similar to be considered as representing the same social reality. The worlds in which different societies live are distinct worlds, not merely the same world with different labels attached. . . . We see and hear and otherwise experience very largely as we do because the language habits of our community predispose certain choices of interpretation.

Nature then becomes a mirror of culture as expressed in language; even the empirical view of nature is a reflection of culture. As Terence Hawkes puts it, "Only the advent of post-Renaissance rational humanism, the invention of 'Man' as an entity separate from nature, concerned to operate logically *on* it, not cooperate analogically *with* it, conceals that from us."

We have the same problems in knowing the self that we have in knowing nature. Even the virgin territory of the unconscious—that inner wild place—can only be understood through the veil of language. Jacques Lacan, a controversial French psychoanalyst, writes in *The Language of the Self*:

Whether it sees itself as an instrument of healing, of formation, or of exploration in depth, psychoanalysis has only a single intermediary: the patient's Word.

There is a danger in not fully realizing the mediating function of the word, however; a person ends up with what Lacan calls the "empty word": An attempt to know the truth of the self is destined to fail because the language of that knowledge will always be supplied from outside, from the community. The unconscious forever recedes from the words that would hold it, always in some measure hovers just beyond full grasp, and it is a paradox to say it can be known. To misread the empty word is to accept a mask for the self, a delusion for the reality, a distortion for a window.

It may be that we can know the earth, people, and even ourselves only as we invent them, only as they live in our words and dreams. Everyone's parrot is necessarily filtered into his or her life through the distortions of will and desire.

Yet this alienation, it seems to me, has its limits. It too lives only in the mind.

The gun shoved in my face is an eloquent refutation of all this philosophical theorizing. The parrot in the tree is another, not violent like the gun but beautiful and alive. Because, in a way that is ambiguous and constantly shifting, I am involved in the world by the very act of perception.

Using psychoanalytic terms, Lacan describes the ambiguities of knowing the self through what he calls the "full word," that is, living with the knowledge that the self is constantly being recreated as it is summoned up in the "hysterical revelation." The self gets located in "the Imaginary and the Real."

The crazy thing, what really scares me, is that these paradoxes make sense to me. They are the sign, themselves, of the pointlessness, the inadequacy, of our division of the world into real and imaginary, the split of nature into outer world and inner consciousness. For Lacan, we can summon ourselves into language and yet at the same time find ourselves slipping away. The drama of knowing the world, of knowing ourselves, is a perpetual reenactment and re-creation, always both true and false.

In all these paradoxes, perhaps we've reached the horizons of what can be thought with our centuries-old division of the world into same and other.

Still, even if I can't know my world completely, finally, absolutely, that's no excuse for not living in it. That doesn't mean I shouldn't engage in the politics of being. I want to locate myself as fully as possible in the world. I want to be right there, to feel the storm, to break the silence, to see too much, to live beyond my broken dreams. That means confrontation sometimes, being willing to declare myself and challenge the illusions of the moment. And our ideas have consequences: Endangered animals have suffered because of the way we see them; it is not enough to say that we cannot know them completely, that one view is as good as the next. We have to testify to what we've done to parrots and to what we see in our lives. This means facing up to the pains of history and admitting what we're doing still to thousands of other animals.

I also want to know, sometimes with incredible fervor, what goes on inside the mind of another person or another creature.

Confrontation and conversation—or, to put them in terms of the animals I've talked about so far, the howling rhetoric of a wolf or the quieter dialogue of the manatee. In either case, we have to be willing to acknowledge the limits of our vision: We must scrutinize them, test their boundaries, expand them, and even destroy them. I'm willing to risk both confrontation and conversation to break down the barriers between who I am and what I can be.

I find myself someplace between the gun in my face and the parrot in the tree, between the wilderness that has no meaning and a feather that flashes in my hand. In the interior ecology that I carry with me, expressed in words and images and symbols, everything that I have said and done and seen is somehow connected. In this inner world, the boundaries between the human and the animal sometimes grow ominous, sometimes shimmer, and sometimes magically dissolve.

VIII

Toward the late afternoon, as I sat on the club moss, an anolis lizard suddenly appeared on a tree trunk about 5 feet away. It snapped so quickly into place, stopped so sharply, that it seemed simply to materialize on the tree. Head pointed downward, tail running up the tree, it stood on its four feet, spread wide, and began to bob up and down as if it were doing push-ups. I have no idea why. Then it flared the orange dewlap under its chin, a round proud blaze of color on an olive lizard. I took it as a proclamation: "I do exist."

Against my back, I was leaning into a *Clusia rosa*, a tree with round, wide leaves, thick as leather, shaped like teardrops. The cushion of club moss beneath me oozed luxuriantly all around the tree, like green and leafy lava. Club moss is one of the most ancient plants. On the ground, ferns of several sorts, still wet from the rain, unfurled their voluted heads as if you could watch them opening up, their delicate fronds glistening with the fresh green of new wheat. Other trees, too, were all around: the caimitillo verde, a hardwood, with its very common, flat oval leaves; the sierra palm, with its stilleto leaves vibrating in the soft winds. I began to see in all these leaves of the forest the outward expression of rich vitality, happening now, unified and coming alive in the dishabille of its varied greenness, khaki and olive and bright emerald.

I felt like the trees all around me: strong, but without deep roots.

I waited for the parrots to return, and I thought about being alive. Not far from his house, Wayne had shown me a small white cross with the name Rodrigues on it. A few months earlier, this man had been abducted from the airport, brought into the forest, robbed, and shot, and his body had been dumped. I wondered why the Puerto Rican with the gun had not shot me, and I thought about what must have been going through his mind as he led me down the road, before pulling his gun.

Just as Wayne had predicted, at exactly four in the afternoon I was jolted out of my thoughts. The parrots had come back to the roost, announcing their return in a riot of flight calls, a hee-haw, hee-haw that sawed through the immense stillness of the forest. Through small breaks in the canopy, I could see several parrots circling against a China-blue sky, all the while clamoring in a wild, agitated uproar.

Two of the parrots had landed high in the tree above me. Either they were in a duet, answering each other back and forth, or they were calling to the parrots still pounding around in the sky. I craned and peered and scanned the branches with my binoculars. They called a few more times. When I finally saw one of them, about 50 feet up in the canopy, perched on a bare limb, the parrot was looking down on me, cool and intent.

One of the eerie feelings I often have in the woods is that animals are out there seeing more than I do and usually seeing me long before I see them.

About a foot long, the parrot was almost solid green, with darker edges to some of the feathers on its body, giving a scaled, roughened effect. In the strong afternoon light, it was clear that, for all the power of its voice, this parrot had to be known by its subtleties—the shaded, mossy greens of its body; the hints of turquoise blue under the tail; the softened hues of chartreuse on its shoulders and neck, much like the yellow-and-green harmonies on a ripening lime.

Preening, the parrot bent over and stuck its head under a wing, which spread like a fan of blue satin opening against a paler blue sky.

Short-necked and cheeky, its big head looked even more ponderous with its massive, pale beak. And on its forehead, a blaze of red, like a flame decal on a hot car.

The wind swayed the branch softly, and one of the parrot's tail feathers blew askance with the same casual negligence of a strand of hair that falls from a woman's head and dangles in a breeze.

The parrot kept silent, rocking, feet gripping the limb, shoulders leading its body through slow arcs. It moved with the fluid, slow-motion definiteness of a mime, both graceful and halting, stepping sideways on the branch in deliberate, stealthy glissades. The bird's motion reminded me of the leaf-flow of the trees in the wind.

What was most startling, disconcerting even, was the expression in the parrot's eyes. I was reminded of the look I saw when I was flying with the condor, but this was different: I was not in the skies but on the ground. With the parrot, I felt less like I saw it than that it saw me. I could see its eyes, dark and liquid, moving as they looked at me, searching and alert, taking in everything. Since the parrot's eyes were surrounded by a brilliant foil of white, its gaze seemed exaggerated, imposing, intimidating even. It followed me in its gaze, cocking its head, watching me.

Even its beak had a bemused and slightly ironic grin to it that left me feeling somehow condescended to.

Earlier that morning, I had seen parrots, but the experience was nothing like this. Then I had wanted to rise into some sanctuary of unspoiled life, beyond the reach of guns or the degradation of an island. Like the transcendentalism of Thoreau and Emerson, which continues to inform much of our understanding of nature, the morning parrots suddenly seemed like only a pastoral escape from the anguishes of life. Much of our view of nature is a kind of monasticism cloaked in green, a love of solitude and purity. The danger of this romantic escapism is the contempt it breeds for humanity.

But if we are honest, most of us will realize that we *do* turn to nature as a way out of the squalor of history and the pain and triviality of our lives, as a place to pursue our own dreams and the echoes of our selves. It's unavoidable in some measure, I suppose, like Narcissus in the Roman myth, chasing after Echo. In the modern world, we are not so much alienated from desire as confined inside it, blindly or willfully pursuing it even across the glittering theater of nature.

In *Tristes Tropiques*, Claude Levi-Strauss gives his interpretation of the way the Bororo tribe of South America understands their relation with the brilliant red parrot, the macaw. The macaw, Levi-Strauss claims, is not part of a separate world but part of the human world: "Thus the Bororo believe that their human form is a transitional state: between that of a fish (whose name they have taken as their own) and that of a macaw (in the guise of which they will finish their cycle of transmigrations)." In their division between nature and culture, animals are in the world of people. Their identities are enmeshed in nature. The Bororo are not *like* the macaw. The Bororo, according to Levi-Strauss, are in the process of *becoming* the macaw.

The Puerto Rican parrot stayed on the branch, preening and watching me, for the better part of an hour.

I no longer needed a refuge. This forest was a place: I was here, and the parrot was here.

From its perch, the parrot looked down on me, and I did not rise above the world, or above other people, or above myself. Its stare stripped me and pinned me to my own life and to life itself. It is a rare experience for me, always an achievement, to feel myself becoming visible, taking shape, assuming a place. I felt present and alive.

The more I felt, and the larger my heart got, the more too that I hurt with a pain connected to the love. My expansiveness was indiscriminate, taking in people as well as parrots. I had a rush of love for Wayne Arendt and for Gilberto Rodriguez and for Nina

Cramer, and the others who had helped me, as people had helped the parrot. Affection cut with sadness—for what we do to each other, for what we give to each other. For the simple outlines of lives whose stories are told in crisis and precarious recovery.

The parrot finally flew off its perch, honking along with the other one that had been in the tree, and together they joined nine other parrots above the canopy. In one last gregarious fling before settling onto the roost for the night, the flock squawked its way higher into the sky, flying in circles that slowly widened. The slanting evening light glanced off their green-gold bodies and fluttering blue wings. Against the depthless blue of the sky, they were tiny iridescent splinters of life.

SEVEN

GREEN EYES AT NIGHT

I

Brilliant as emeralds, big as headlights, breathtaking in their effect. Green eyes shining at me in the night.

I froze.

On my back, I wore a 45-pound pack, filled with a 12-volt, jell-cell battery, which powered an enormous spotlight in my right hand. Not wanting to blow one of the rarest moments in my life, I grabbed my binoculars, dangling around my neck, with my left hand. And in a trick of dexterity that's almost absurdly impossible, I tried to hold the binoculars to my eyes with one hand while I aimed the spotlight at the green eyes in the dark with the other hand. It wasn't so much a balancing act as a test of coordination, which I failed miserably, the spotlight waving stupidly all over the darkened prairie, my binoculars pointed who-knows-where into the night.

Cautiously, keeping my eyes fixed on the spot where I'd seen the fierce green shine, I shuffled forward, hoping to close in, get a better look, try to identify what I was seeing.

Two steps and crash.

I walked blindly into a prairie-dog burrow. The bottom fell out from under me as if I were on a circus ride. I lost my footing and tumbled to the ground, relaxing as I dropped under the weight of the pack, giving in to the fall to avoid breaking a leg, forgetting the laser-green eyes for the whole ridiculous descent, comical as the night's fool.

One of my chief joys in looking for wildlife is the exhilarating, ambiguous time between first sight and final recognition. For a delicious and crazy instant, anything is possible. It's one reason I love looking for nocturnal animals. In the darkness and shadows and silhouettes, I'm groping for sight, tripping over my own blindness. Searching the shapes in the darkness, I'm looking as much with my mind as with my eyes. The wilderness at night is a darkened stage where I constantly surprise myself, as if I'm both actor and author of my own dramatic ironies. It reminds me of Bottom in *A Midsummer Night's Dream*, who for one wonderful night becomes a beloved ass for the Queen of the Fairies: At night in the wilderness, with a little luck, every Bottom's ass has a chance for his own bottomless dream.

Scrambling unceremoniously back to my feet, I scanned the darkness with the spotlight's intense beam and picked up an indistinct shape bounding across the penumbral landscape of the prairie. Sleek and agile, its back rolled like the curl of a wave making toward the shore, its tail undulating behind like a second, succeeding wave. Fleeing but curious, it would halt abruptly and look nervously back at me, head bobbing, its eyes charged with energy in the reflected light of the beam. An animal seen but not seen. Luminous green eyes shining out of the shadows.

I was hoping this was a black-footed ferret. Its scientific name is *Mustela nigripes*. To me, the black-footed ferret is the perfect symbol of an animal as mystery, and of our own groping for vision in the dark, with only narrow beams to help us see. Though its range once extended across twelve western states and two Canadian provinces on the Great Plains, the black-footed ferret has always been elusive, secretive, and rarely seen. It's an underground animal, living most of its life in prairie-dog burrows. And it's largely nocturnal, emerging in forays at night to hunt or move to another burrow. A small animal, about 1½ feet long, this member of the weasel family had been feared extinct at least twice in this century. Slipping in and out of our awareness, it was thought to have vanished nearly as often as it had been known to exist.

It has been, as well, the least studied and the most poorly known of all the major mammals in North America. Everything about the black-footed ferret—the difficulties in managing it, its elusive life on the borders of our awareness—suggest how superficially we have penetrated the mysteries of the world around us.

At the time of my visit to the prairie, early in August 1986, the black-footed ferret was dragging itself and everyone associated with it through the most traumatic wildlife catastrophe in North America in this century. We knew the creature was out there, but everything else—how many there were, what was wrong, whether the species would survive—was uncertain. Just two years earlier, in 1984, managers of the only known population of black-footed ferrets had been swelling with optimism. This colony of ferrets was in the dusty scrub prairie along the Greybull River and the Absorkas Mountains, near the small town of Meeteetse, in northeastern Wyoming. In the ranchlands south of Cody, ferrets had been discovered in 1981, and for several years the population burgeoned. In 1984 the official estimate, based on nocturnal surveys with spotlights, was 128 ferrets, an unprecedented number.

Biologists had worries, though, especially over the possibility of some natural disaster. Nevertheless, the colony was breeding, growing, and expanding its range.

The surveys in 1985, however, brought the worst possible news. Sightings of ferrets were way down, and it turned out that canine distemper had run an epidemic course through the colony. Several ferrets brought into captivity developed the disease. They grew red spots, their skin wrinkled, and they died in fevers.

Panic spread among researchers. In national magazines, the local "hicks" of the Wyoming Department of Game and Fish were accused of mismanaging a national treasure. This was turning out to be, they said, the first time an endangered species had been managed into extinction. In an emergency move, Wyoming Game and Fish decided in October to capture every remaining ferret they could find in the main colony.

They found only six.

The next year was a winter of hell. The six ferrets in captivity did not breed, and no one knew if any ferrets had survived in the wild. For conservation biologists, the black-footed ferret had become the ultimate horror. Bitter charges of incompetence appeared in national print, the principle one being that Wyoming Game and Fish should have started a captive program much sooner, long before the crisis hit. The fight was reminiscent of the brutal conflicts over the California condor, in both its virulence and its terms: captive animals versus wild populations. But it was much worse, really, because the paroxysm of loss came so suddenly, in just three months, and the tragedy of death and disease was so much more widespread.

The overwhelming feeling of doom that winter led many people to predict in print that the ferrets were now extinct in the wild. The real point, though, was that nobody knew what was going on with the ferrets—how many, if any, survived and whether the ferrets had any future.

When I joined the survey crews the next summer, in 1986, the black-footed ferret was without question the most endangered and the most controversial mammal in the United States. Dave Belitsky, of Wyoming Game and Fish, coordinated all the agencies working on the black-footed ferret, and I joined him in scouring the gently rolling plains for survivors, night after night, dusk to dawn.

An unpretentious man, Dave had borne many of the allegations of incompetence; his main way of dealing with them was to keep his faith in the wild ferrets. His career had already included work on the endangered Kirtland's warbler, as well as the Puerto Rican parrot. He was convinced that there had to be ferrets out there in the night. Not just surviving, but baffling human predictions of their extinction, as they had earlier in this century. It seemed to me—though this was only speculation—that Dave had private reasons for hoping to find wild ferrets that summer. The surviving wild ferrets would help redeem whatever errors he and others had made in managing them.

I had been helping Dave and his crew of researchers look for ferrets around Meeteetse for well over a week when I saw those brilliant green eyes. I still had not seen a wild ferret. The importance of identifying these green eyes was not, however, in proving that ferrets still survived in the wild around Meeteetse. Some of the pressures of the last winter had already been alleviated for Dave and his crew of spotlighters: In the last two weeks, they had managed to turn up several ferrets. By no means had they found the prolific abundance of former years—that was a thing of the past. But they had found some vindication: twelve wild ferrets—including two females with kits.

Ferrets were still breeding in the wild. These numbers didn't even come close to taking the ferrets out of jeopardy, and they represented small consolation for managers who had devoted the most important part of their professional lives to the ferrets. Nevertheless, twelve ferrets constituted at least a glimmer of hope and a measure of relief.

The problem was that one of the two families had disappeared—a mother and three kits. For several nights, Dave and I had kept constant vigil by the den, sitting in a truck, watching the family's burrow through the small hours, waiting for the mother to appear. We scanned the prairie-dog town with huge searchlights mounted to the truck, trying to find her on the hunt. It was crucial that the family be all right.

All we got for our pains were empty views of a prairie-dog town, looking increasingly morbid and funereal, mocking our spotlights. The prairie-dog burrows, dug into the treeless earth, were small craters scattered through a barren field. As the nights passed, the scene looked increasingly to me like a field of dug-up graves; I felt that we were at a wake. Between the crash the previous winter and the disappearance of this family, it seemed that ferrets were always living on the edge of their own destruction.

Dave stayed calm, or at least his stoic composure betrayed only hints of anxiety. For Dave, ferrets were less a question of knowledge and control than a question of faith. He believed in the wild ferrets and their ability to survive on their own, which is partly why he had not advocated a program of captive breeding earlier. This faith had already gotten him into trouble. Yet every night, as we watched from the truck, he was convinced that they'd turn up. Ferrets, he averred, could spend several nights in a row underground. But by the eighth night, even Dave was showing nerves, smoking his pipe more frequently, talking more about where the ferrets could be.

I had my own reasons for needing to find these ferrets, in addition to my concern for the lives of this family. Like Dave, I had been living through my own hellish season.

I had been collapsing beneath my own contradictions. The richness of response I had felt in Puerto Rico, with the parrots, was gone. Since then, I had been to two funerals, and I had been sick. I had lain in bed for days, exhausted and weak. But these were only symptoms, mere hints at a deeper, internal desolation. My marriage was careening erratically but inevitably toward divorce, through discoveries of betrayal I had never thought possible, leaving me staring once more at a bleak wilderness within. I had just moved out of the house for the third, or maybe it was the fourth, time.

The ferrets put me in touch with the sorrow it feel for all vanishing species: I'm searching for something, but I'm not even sure as I'm looking what it is that's been lost.

From time to time, Dave or I would get out of the truck with a big searchlight and hike through the prairie-dog town looking for the ferret. That's what I was doing on the night when I stumbled into the burrow.

Once on my feet again, I chased eagerly but discreetly behind the dark shadow with the green eyes, hoping that it might be a ferret. It might not be a ferret. I had already had many false rushes. Other animals on the prairie have green eye shine: pronghorn antelopes, though their size makes them easy to distinguish from ferrets; skunks, stockier than ferrets in their silhouettes, but similarly low-slung; and weasels, smaller than ferrets, but the most likely to be mistaken for them.

As I ran, vesper sparrows flitted out of their beds under the sage and short grasses. Several cottontail rabbits skitted before me, their eyes reflecting blood red in the spotlight.

The creature with the green eyes lured me through a small draw and across a rise. It stopped along the margins of an irrigated field.

I stopped too. The clayey soil puffed around my boots in dry clouds. Jupiter shone near the horizon, ice blue. The moon was just rising, luminous in its own distant, desolate beauty. Beyond the green eyes I had been chasing, a hayfield lay in darker shadows. Green eyes glared out there, too, at least seven pairs, like floating phantoms in a weird sea: mule deer, grazing on a rancher's hay.

I put the spotlight back on the green eyes I had been chasing. I was shaking with excitement, because I knew the creature might be the mother ferret. Jumping up onto a culvert, it watched me for several minutes. Then, skipping to the ground, it rippled along margins of the field, darkly supple and buoyant in its gait, turning occasionally to stare at me with those otherworldly eyes.

I never saw it distinctly before it vanished, but I knew—I just knew—it was a ferret. I started back to the truck, skipping and jumping with my pack on my back, rushing to tell Dave what I'd seen. I had not identified it with anything approximating the empirical certainty we needed, and we'd have to find it again to confirm the sighting. But in my gut I felt sure it was the mother ferret we had been looking for. Ever since I had become interested in wildlife, as a graduate student about fifteen years earlier in Minnesota, I had longed to see a black-footed ferret. I had seen its picture in field guides to mammals, I knew it was extremely rare and almost extinct, and I had let it become an object of strange desire. I had always thought it would be unbelievably wonderful to see one.

All those years crowded into this moment—seeing those brilliant eyes—and all my recent losses, and all my worries over the ferrets. I hurried back to the truck, drunk on the vision of laser-green eyes and that shadowy, barely seen body.

II

The next morning didn't simply arrive for us on the prairie. It seduced us. In the first gray adumbrations of dawn, coyotes howled behind the nearby hills. The faintest of blues brushed the tops of the hills, pushing against the black sky, and orange snuck up behind the pastels of blue as the light gained intensity all around us. To the south, flat-topped benches of ancient rock stood out stark as skulls against the long, rolling lines of the plains, their faces abraded by centuries of wind, exposing the hard-rock below. At their base lay talus slopes of fallen rock, like last night's scattered clothes around a bed. Their granite tops caught the first light in glowing saffron. The open spaces of the prairie filled with honeyed light, which glittered off the dew-wet sage. The sky became as brilliant and moist as a split cantaloupe, with clouds shaded in purple like the voluptuous curves of eggplants.

The morning stole upon Dave Belitsky and me like a memory —the kind that begins with a single phrase from a forgotten song, reminiscent of an old love: You slowly give yourself over to the past, filling with remembered feelings, and then without knowing how it happened you realize you're inside the memory.

When I ran back to Dave the night before, full of enthusiasm for my green-eyed discovery, I roused him from a near sleep. In the truck, we moved to a slope overlooking the field where I'd seen the eyes disappear, hoping to confirm a ferret. Big-bodied and baby-faced, Dave stayed cool, battle-tested by his years of ferret work. He puffed his pipe and tempered my excitement with a well-prac-ticed, laconic realism: "Could be a weasel."

It didn't take Dave long, though, to spot what turned out to be the mother ferret, prowling through the gray sage among the prairie-dog burrows. Low-slung and swaybacked, she charged along the ground through the sage, vanished for several minutes, and then popped up on her haunches like a periscope above waves, surveying the prairie with stealthy glances. Slender black legs and feet dangled down her tawny belly. A sleek neck, fully a third of her total body length, stretched in elegant hues of cream and silver, contrasting with her charcoal chest. Her face was alive with a cat's alert concentration: Rounded feline ears crowned a broad forehead, which slanted in a sharp triangle to a pointed snout with a button nose.

Across her face was the characteristic marking of her species, the pure sign of her poetry and her mystique: an outlaw's black mask.

We have never known much about the ferret. It has been so elusive that we've only had a few, brief chances to study it. Before 1940, biologists had almost no records of the black-footed ferret. It was probably never abundant. Throughout its vast historical range —up to 100 million acres of short-grass prairies on the plains—the ferret was closely associated with its main, and perhaps only, prey, the prairie dog. But, despite the numbers of explorers on horseback and pioneers in wagon trains that crossed the prairies, the ferret was not recorded by white men until 1851, when John James Audubon and the Reverend John Bachman announced on the basis of a trapper's tattered skin that they had discovered a new species.

That specimen was lost or destroyed, and some naturalists came to doubt Audubon's description (he was known for his extravagance and his love of finding new species). It took another quarter of a century before Dr. Elliot Coues was able to locate additional skins and confirm the ferret's existence.

For the next century, the ferret remained largely a memory. However, its dependence on prairie dogs during that period was probably responsible for its current endangered status. The fate of the more celebrated bison on the rich pastures of the Great Plains is well known: According to Peter Matthiessen in *Wildlife in America*, millions of bison, once forming the largest herds of animals on the face of the earth, were slaughtered within a generation. At about the same time, the black-footed ferret was beginning a much less heralded course toward extinction.

In the wake of the first explorers and scouts, and once the Indians had been squashed in the 1880s, the livestock industry grew. Ranchers came to view the prairie dog as a competitor for the limited forage of grama grasses. From the 1920s to the 1960s, the federal government sponsored intensive programs to eradicate the prairie dog throughout the Great Plains states, pumping poison gases into millions of burrows and leaving out grain loaded with toxic chemicals. Thus the ferret lost its main prey: Today there is only a small fraction of the number of prairie dogs estimated for the late 1800s. Some states have eliminated 99 percent of these rodents. The gases probably also killed thousands of ferrets outright. And plowing of the grasslands consumed the ferret's habitat.

By the 1950s, many people considered the ferret extinct, and it had never been carefully studied.

But on August 7, 1964, William Pullins was checking prairie-dog colonies in Mellette County, South Dakota, for the U.S. Bureau of Sport and Fisheries. His job was to control, by gun, poison, or trap, animals considered obnoxious, and he was sizing up a poisoning job on a town of prairie dogs. In the process, he saw a black-footed ferret, a species he had seen no more than a dozen times in his life. Firm sightings of the black-footed ferret were so rare that when word reached F. Robert Henderson, a biologist for the South Dakota Department of Game, he vowed to study it. Henderson's study culminated in a large monograph in 1969. Meanwhile, a graduate student at South Dakota State University, Conrad Hillman, undertook the study of twenty-one ferrets in six different prairie-dog towns in Mellette County, a study which became his master's thesis. While not definitive, both studies remain the seminal works on the ferret.

Then, in 1974, the ferrets in Mellette County simply disappeared. No one knows what happened.

Over the next several years, government funds to study or search for ferrets dried up, and more and more biologists were grimly coming to think the ferret was extinct. The U.S. Fish and Wildlife Service was ready to agree.

However, on September 26, 1981, a friendly, black and white ranch dog named Shep made a startling contribution to science. In the middle of the night, in a loud commotion, the blue-heeler crossbreed killed a strange-looking mammal on the porch of his owners, John and Lucille Hogg, just outside Meeteetse, Wyoming. Not knowing what it was, the Hoggs took the body to their local taxidermist, LaFrenchie's, who confirmed that Shep's trophy was the nearly extinct black-footed ferret. The rediscovery of the mysterious ferret had come with incontrovertible evidence—a corpse.

Biologists searched the area, found ferrets, and soon were studying a thriving colony.

The folk wisdom around Meeteetse, given the subsequent extirpation of the ferrets in this colony during the mid-1980s, is that the Hoggs should never have reported their discovery to the government. Though the numbers of ferrets climbed for three years, and though canine distemper was clearly the immediate cause of the horrible crash in the local population, the Hoggs are convinced that the real cause of the ferret's epidemic was the hordes of government officials and biologists who swarmed into Meeteetse—looking for ferrets, studying them, trying to manage a population that previously survived without their help.

When John originally found the dead ferret on his porch, he tossed it over his fence. "That's where it shoulda stayed," Lucille said once to me. "Cuz then maybe them pore little things woulda lived. I think it's stress, all them floodlights and radio collars and traps. If anyone sees another ferret now, they're not gonna say anything. I wouldn't. If one crossed my porch again, nobody'd ever know."

Ironically, Dave Belitsky, who coordinated the agencies and biologists working on the ferret around Meeteetse, finds justice in the Hoggs' attitude. The loss of the ferrets is a pain he has to live with, for he, too, feels that the research itself may have harmed them. As Dave said to me, even the distemper may well have come from the researchers, carried into the field from a family pet.

In addition, Dave told me that during the summers of 1985 and 1986, most adult ferrets in the Meeteetse colony were trapped, anesthetized, and marked with ear tags or fitted with radio collars (and sometimes both) at least twice. The competition between federal, state, and private organizations in managing the ferrets was intense. A premier endangered species like the ferret offers biologists the chance of a lifetime for making a career, so the drive for "data" on these ferrets may have been spurred by more than the desire for information that would lead to more effective management. "Have you noticed," Dave asked me, "that recent photographs of Meeteetse ferrets are of animals with their ears tattered from tagging?"

The opposite point of view, however, is that managers should have intervened more forcefully, especially in establishing a captive colony much sooner, when the populations were strong and healthy, as insurance against just the sort of catastrophe that did happen. With hindsight, Dave agreed that they'd made a mistake in not acting sooner. Hands off versus hands on, the wild ferret versus a more intensively managed ferret.

Even as we watched the mother ferret prowling through the prairie, there were serious discussions in progress over whether to capture these final remaining wild ferrets in Meeteetse. Dave opposed capturing them all. Though the matter was still officially unresolved at the time, I knew that she would soon be captured. The precedent of the California condor, the sheer weight of the final year's horrible losses, plus the less tangible but even more powerful imperative of the problem-solving mentality (we have to *do* something)—they made the decision to capture the ferrets, in my mind, inevitable.

The prospect of the ferrets being captured added a bitter poignancy to the morning. The following fall and winter, all twelve wild ferrets were in fact caught in traps and placed in captivity, bringing the total population to eighteen. All in cages.

The prospects of captive breeding? At that time, in 1986, they were completely uncertain.

In the mid-1970s, the Patuxent Wildlife Research Center, part of the U.S. Fish and Wildlife Service, had tried in vain to breed captive ferrets. Working with nine ferrets, biologists managed to breed their only compatible pair three times. Two of the matings produced babies that died shortly after birth. The third year yielded a false pregnancy.

The captive-breeding program in Wyoming has turned out to be unexpectedly successful, radically changing the mood of the ferret recovery project. Working for Wyoming Game and Fish, the chief veterinarian on the project, Tom Thorne, achieved a major breakthrough in 1987: For the first time, ferrets were bred, born, and successfully raised in captivity. Though ferrets are violent when they mate—it's basically a rape, according to Thorne—seven surviving babies were produced. Then, in 1988, came another breakthrough: Forty-four ferrets were born in captivity, and thirty-four survived.

As of summer 1988, the original captive population of eighteen ferrets had jumped to fifty-eight. Tom Thorne is so optimistic that he hopes to have 200 to 250 captive ferrets by 1991, which is when Wyoming Game and Fish is now planning to release ferrets into the prairie-dog towns around Meeteetse.

Dave and I knew none of this, of course, as we watched the mother ferret. Nor did we know the matriarchal importance she would assume in captivity. Named Jenny after she was captured, this mother came in from the wild with three young, one a daughter that was given the name Becky. In 1987, the first year that ferrets were successfully raised in captivity, Jenny and Becky produced all seven of the surviving babies.

One outstanding question concerning ferrets remains: In their huge range, with all the prairie-dog towns between Texas and Canada, are there still undiscovered colonies of ferrets? If the ferret experience adds impetus to the strategy of using captive breeding to save severely endangered animals, it has also had another effect. In the enthusiasm engendered by the baby ferrets, searches for wild ferrets have begun again. The ferret has become, once again, an object of desire.

Despite the success with captive breeding, the underlying conflict in the management of ferrets has not been resolved. Do we trust wild populations to survive with a minimum of management, or do we undertake active intervention? Do we agree with the Hoggs, for example, that wild ferrets should go unreported if they are found, in order to protect them from researchers and government agents? Or do we agree with Tom Thorne, who hopes the disasters of the Meeteetse colony will not prevent people from reporting ferrets they might see? As he said to me about the Meeteetse colony, "If the Hoggs had not reported their ferret, the colony might have gone extinct anyway, especially if the canine distemper came from natural sources. There would have been no opportunity to expand the biological knowledge through field studies or, more important, to attempt captive breeding."

As Dave and I watched the mother ferret, though, the final captures and the successes with captive breeding were still in the future. The ferret raced among the prairie burrows, peering in, sniffing, running to the next one with her black-tipped tail bouncing behind. All energy and grace, she was a sort of visual kinesis.

She was probably hunting. Prairie dogs are as big as ferrets, and they have sharp incisors. They also sleep at night, when ferrets are usually hunting, and hibernate during the winter. So one theory is that ferrets kill most of their prairie dogs when the prey is sleeping in the burrows.

By 6 a.m., the ferret had disappeared behind a small rise. Dave walked carefully out onto the prairie, beckoning me to follow. She had probably moved her litter to this new area, and he wanted to see if he could find her in one of the holes. The sun was now fully up, hot on my shoulders, and the sage smelled sweet as it scraped my Levi's. Just ahead, Dave put his fingers to his lips and waved me toward him.

I climbed the crest of a large knoll with a prairie-dog burrow in it, bleached dirt packed hard and dry. Down inside, where the mouth narrowed to a small tunnel, the ferret's masked face peeked out at us, 5 feet away. She shuffled daringly out of the burrow, right up to my boots, stuck her sleek neck out, and yipped defiantly. Then, in a quick retreat, she scooted backward into her burrow.

She seemed both bold and shy at the same time. Her confrontations with us, creatures so much larger than she, were almost comical in their tentative audacity, and touching in their frightened intimacy. From the cover of her burrow, her masked face stared out at us, dark eyes glistening with a carnivore's vitality. She hissed like a cat, yelped like a small dog, and rattled at us in crackling bursts, like high-voltage static. Several more times, she charged up to my boots, yipped, and fled.

She reminded me of myself: how I love to show myself, and then, horrified to find myself out in the open, scurry back for cover.

Before white men discovered the ferret, the Plains Indians knew of it. The Crows, for example, were aware of its rarity and attached mystical significance to its masked face. Animals were a source of "medicine," or power, communicated through dreams and visions. The nature of that power, I have been told by Indians I've talked with, cannot be explained theoretically or understood from the outside: It is an extremely personal relation between the individual and the animal. Indians have explained to me that this medicine is spiritual and that it exists in the world, everywhere. Dreams and fasting offer windows of communication to this spirit world. Receiving this medicine requires a different posture toward an animal than whites understand. It is not an arrogance but comes from humility, from recognizing that you are nobody and that nature is full of meaning. A seeker humbles himself, prays to an animal for medicine and vision, and makes himself small to the spirits, who speak through nature or dreams. Relics of this medicine, and of the animals' roles in conferring it, can be seen in the animal bundles that Indians made. Several pouches I saw were made out of ferret skins, wrapped with bright yarns and decorated with beads and feathers. If you are lucky, Indians have told me, an animal may take pity on you. It may act as a messenger of the spirit world, conferring upon you a "medicine," a power or a path.

I have never known this power, not from personal experience. But I have known that moment of stripping away, where I am little more than a vessel to be filled. And I have felt intangible strength from the presence of animals.

I watched the mother ferret barking at me from her burrow, and I felt absolutely sure that more ferrets survived in other prairie-dog towns. Though I hoped others would be found on searches— hoped that federal money would be pumped into the effort to find ferrets—it didn't really matter. The numbers we use to define pop-ulations, to tell us what's living and what's not, seemed irrelevant. We are probably living beside them, in a blind proximity. Ferrets very likely survive whether we know it or not, playing hide-and-seek with us.

It was an absurd and abundant confidence, but I knew ferrets were alive out there as surely as I had known, the night before, that those green eyes were this mother ferret. It was something like my faith in the defiant and fragile presumption of life itself.

The mother peeked out of her burrow, head jerking nervously as she barked. I pictured other ferrets, ferrets we might never see, hidden in their burrows on other prairies. I could see them in my mind, peeking out hesitantly into this same dawn—rare, elusive, unpredictable. Not every secret, I thought, has to be seen. Not all mysteries have to be known.

EIGHT

"SO IGNOBLE A LEVIATHAN"

I

For all its infinite reach and scope, you don't really *see* the ocean from the bow of a small boat. You *feel* it. The wilderness offshore tosses itself into your face with the spume of a sickening beauty.

I sat spraddle-legged on the bow of the 29-foot *Harry Lee II*, named for Robert E. Lee's father, as we chugged out of the mouth of the St. Mary's River dividing Florida from Georgia. Squalls from a threatening storm whipped the ocean surface into whitecaps. Low clouds smothered the sea in a sepia wash, with lurid greens mixed in, and waves beat the hull in 5-foot breakers. One of the volunteers puked aft over the railing. My face dripping with salt spray, I rode the boat as if it were a rodeo bull as it crashed through the breaking waves, making our way a few miles offshore. At the shock of each wave, I held on tight and felt the boat shiver through my thighs. The sea as an assault to our senses: I kept thinking of James Joyce's description of the ocean in *Ulysses*: "the snot-green sea, the scrotum-tightening sea."

The solidity of our bodies is an illusion: The body is its own ocean. Though I live with the firmness of my flesh everyday, my body is a *mare incognita* to me. Our skin and bones and muscles are liquid—waters of the body held in fragile membranes—made up of some 65 percent water. And the seas are only apparently water: They are heavy, firm, tangible in their substance when they pound into us, hit us in the face. It must be that this is why I love to live by the sea, why I keep making trips out onto the sea, why I have taken so many chances to go offshore looking in the fecund barrenness of the surface of the sea for albatrosses and petrels, turtles and whales. Some physical pull, some response in the body, exerts its sway over me, and I keep going back, as if the sea is a moon to my tides.

Rocking over the breakers in the *Harry Lee II*, my head and shoulders swung in wild circles while my stomach swam in my guts, and I realized I was answering the sea with my body.

We were looking for whales in the Atlantic Ocean, just off the coast of Amelia Island, at the extreme northeastern corner of Florida. In their size, whales are prodigious creatures, living hyperboles. They answer some need I have to be reduced, to be humbled, to be reminded of how little I am, like Job struck dumb before the invocation of the monster of creation, Leviathan: "Shall not *one* be cast down even at the sight of him?"

Each of the great whales has come to embody a particular meaning for us in the ecological renaissance they helped create. The humpback whales: the most popular of all, probably the current ideal—a "gentle giant." Their eerie and reverberating songs, first recorded in the early 1970s, evoke the lyricism of the sea. And they are one of the most playful of the whales, white flanks shining as they breach. The bowhead whales: associated with Eskimo subsistence hunting and the unending allure for Americans of arctic hardships. The gray whales: famous for their resort to calving grounds in the tropical waters of Baja California and for their long annual migrations up and down the west coast. The orca, or killer, whales: known for their bold black and white markings, their predatory habit of eating salmon, and their place on the totem poles of the coastal Indians of the Pacific Northwest. The fin and sei whales: both rorquals and less well known, but sleek creatures of speed. The narwhals: renowned for their strange and enigmatic tusks, reminiscent of the magic ascribed to the unicorn. The blue whales: over 100 feet long and 160 tons in weight, the biggest creatures ever to have lived on the earth.

And sperm whales: perhaps the most important animals in the American experience. Through Herman Melville's imaginative vision, sperm whales helped create the American identity. They have huge, blunt foreheads and toothed jaws. They prefer the deep seas all around the world. And they have a propensity to ram big whaling ships in anger and to stove smaller whaleboats in a rage of self-defense. Altogether, sperm whales have offered Americans dramatic narratives and noble opponents. They were transformed in whalers' stories into the symbol of our epic aspirations, of our defiance of the "demoniac indifference" of the universe, of our will to conquer the world. They have summarized our vision of the heroic and the tragic in our battles with nature.

But another species of whale has been overlooked, neglected, even spurned: the North Atlantic right whale. Its Latin name is *Eubalaena glacialis*. Its common name indicates its former importance to whalers, who considered it the "right whale" to hunt. Yet its significance—biological, historical, and psychological—has been largely forgotten, even though it was the most important whale in American culture until Melville popularized the sperm whale.

Even if the right whale has been silent to us for over a century, even if we have not heard much about it since Melville's epic, there is something important in this blankness. It was my experience with the right whale off the coast of Florida that first made the significance of its silence come alive for me.

Shrimpers had reported right whales off the coast of Amelia Island that morning, at the mouth of the St. Mary's River, and I was looking for them with biologists and volunteers. Though they were not easy to pick out against the textured but featureless face of the sea, we didn't have to go far to find the two right whales. During a lull in the gathering storm, the seas settled into a wind-fretted chop, and we found the whales cruising along the surface about 5 miles offshore. They plugged slowly into the waves, doing no more than 3 or 4 knots. If we ran the boat at anything above a "fast idle," we punched right past them. In their dark profile, their heads rested partly exposed above the water in an outline like a huge 747 airliner—the same high and rounded shape of the cockpit, sloping up to the blowholes and tapering off to a broad back.

After a couple of hours of gathering data on the whales' breathing patterns—part of a study these researchers were conducting for the New England Aquarium in Boston—we edged in closer, and the whales began to take on dimension and substance. The mother was unwary; as we approached, it became easy to see why, for early natural historians like the Elizabethan Bartholomew, the whale "for greatness of body . . . seemeth an island."

Though not the biggest of the whales, this right-whale mother logged in the water with a staggering bulk, waves washing over her back as if she were shoals. She was about 45 feet long, her back perhaps 12 feet across and very flat. Though her head was long and narrow, she swelled into a monstrous girth just behind her curling mouth. Right whales are the fattest of the whales, with their girth just before their flippers totaling more than half their body length. Their weight in tons is equal to about three times their body length in meters: 15 meters in length, 45 tons in weight. The males have a 12-foot-long penis, and each testis weighs a ton. The right whales also have the smallest brain in proportion to body size of all the whales. The big female near us probably weighed about 45 tons. As we idled up next to her, she seemed massively indifferent to us, her black bulk inert and sulking in the white froth of waves on her flanks. She dwarfed the boat that we were on, and she dwarfed all of us on board, too, as we scurried around to get nearer to her.

It was impossible not to be overwhelmed by her sheer size, by her physical presence. For Western sensibilities, animal life has always been associated with the physical body, and our own bodies are the signs of our animality. But in her the idea of a body had been taken to outrageous and almost transcendent proportions. I stared at her slate-gray flesh and saw some fundamental ground of existence, the space in which life is contained and must live out its time: a living body.

As if her size were not enough of a challenge to comprehension, her face was its own affront. Above a jutting lower jaw, her face was pinched by thick lips that bowed up and inward. On her narrow rostrum (the top of a whale's head from the blowholes to the tip of its snout), she wore crusty, jagged growths of black skin, called callosities, that look vaguely like uncut fingernails curling upward in bizarre shapes. Colonies of "whale lice" clung to these callosities and clustered on her lips, making up patches of white and orange and pink. The lice are little crustaceans called cyamids.

This combination of the fat and the ugly in the right whale has left the American imagination uninspired. Herman Melville, in *Moby Dick*, granted his tribute to the right whale, but it was a grudging tribute, a kind of dismissal or displacement in favor of the nobler sperm whale. Writing of the right whale, Melville invites the reader to "fix your eye upon this strange, crested, comb-like incrustation on the top of the mass—this green, barnacled thing . . . you would take the head for the trunk of some huge oak, with a bird's nest in its crotch." He goes on: "This mighty monster is actually a diademed king of the sea, whose green crown has been put together for him in this marvellous manner. But if this whale be a king, he is a very sulky looking fellow to grace a diadem. Look at that hanging lower lip! what a huge sulk and pout is there! . . . a sulk and a pout that will yield you some 500 gallons of oil and more."

The right whale was slow, found near shore, and buoyant with all its blubber, so it was easy to catch and process. But these same qualities, plus its ugliness, led nineteenth-century whalers to view it with disgust. As the crew of the *Pequod* prepares to lower for a right whale, Melville writes: "All hands commonly disdained the capture of those inferior creatures." And after the crew kills the whale, Melville reports the headsman's attitude: " 'I wonder what the old man [Ahab] wants with this lump of foul lard,' said Stubb, not without some disgust at the thought of having to do with so ignoble a leviathan."

So ignoble a leviathan: given a kind of practical tribute only as the source of immense yields of oil and baleen.

It is not the right whale itself being loathed by the *Pequod*'s crew. It is the right whale as body, with its weird incrustations and its larded bowels, that made our Puritan forefathers, turned Yankee whalers, condemn the right whale even as they killed it. All this immodest bulk and all these strange imperfections of appearance—they are too much of the body. The New England clergyman Cotton Mather has a diary entry from 1700 in which he makes very clear his anxiety over the way his bodily functions connect him too intimately with the beasts:

> I was once emptying the cistern of nature, and making water at the wall. At the same time, there came a dog, who did so too, before me. Thought I; 'What mean and vile things are the children of men. . . . How much do our natural necessities abase us, and place us . . . on the same level with the very dogs!'
>
> My thought proceeded. 'Yet I will be a more noble creature; and at the very time when my natural necessities debase me into the condition of the beast, my spirit shall (I say *at the very time!*) rise and soar.'

So the redoubtable divine resolved in the future to make peeing "an opportunity of shaping in my mind some holy, noble, divine thought."

Captain Ahab transmutes the attempt to deny nature as expressed in the body, transforming a Puritan anxiety into a heroic monomania. His is one side of the essentially modern quest—to travel beyond life, to define the self by throwing it outward through the universe, to tear through the limits of the body and the masks of life. For Ahab, the sperm whale is the embodiment, the great theme and metaphor, of the quest for the self accomplished through metaphysical heroics, of breaking through the walls of the physical to the ultimate and the transcendental, of writing the self through action upon the face of the unknowable. As Ahab says, "All visible objects, man, are but as pasteboard masks. But in each event—in the living act, the undoubted deed—there, some unknown but still reasoning thing puts forth the mouldings of its features from behind the unreasoning mask. If man will strike, strike through the mask! How can a prisoner reach outside except by thrusting through the wall? To me, the white whale is that wall, shoved near to me.

Sometimes I think there's nought beyond. . . . That inscrutable thing is chiefly what I hate."

It was under this metaphysic of the sperm whale that the right whale was lost.

The right whale announces a different side, a different aspect of the modern quest. The sperm whale has a huge square head, inspiring thoughts of the mind-in-nature, but the right whale gives the figure of the massive body. Unlike Moby Dick, the right whale is found not in the "midmost ocean of the world" but closer to home; it is a creature not of far seas but of shorelines.

The white sperm whale Moby Dick, which Ahab chased around the world, implies boundlessness; the right whale suggests the quest for historical roots. It is the story of origins: of the origins of Western whaling by the Basques in the Bay of Biscay and of the origins of American whaling along the shores of New England.

In the silence of the right whale, there is a quieter story, a more humble story, less pretentious but closer to home, anchored to where we actually live. Through *Moby Dick*, the sperm whale locates us in the grand struggle, in the quest for the future, in the impulse to overcome nature—which is so central to the American experience of its own identity. It is as if we are in nature but somehow always moving through nature. But the right whale recalls us backward, into our own history; it locates us not outside of time but in the forgotten past.

There is as well a psychological correlative to this cultural theme. The right whale suggests not just the shorelines of the American continent but the shoreline of the body, which is the ground and background of our lives in nature. It is through the experience of my body that I have my own place in time. It inserts me into the chronology of things. It gives me the space I occupy in the world. And if I am able to discover myself through animals, this is probably their most basic revelation: They summon me to a simple contact with a being which just is—this stupid body, this animal life, which is the basis of desire.

We stayed with the two right whales for the whole afternoon. The captain killed the engine and we drifted with them, chatting among ourselves, taking notes on their behavior, rocking with the swells, enjoying being near the whales. The flat outline of Cumberland Island, off the coast of Georgia, hovered in the distance, through a foggy mist. Gannets flew past us in the poor light, their wings thin and pointed, their beaks thick and heavy, giving their darkened shapes a look reminiscent of ancient pterodactyls. The mother whale loomed beside us, passive for the most part, resting.

Her baby, however, was coltish and frisky, playing around its mother with a surprising, winsome grace. It rested its head across its mother's snout and then slid slowly backward into the water, bubbles gurgling up around it as it vanished. It reemerged in a headstand, tail first. Its broad, flat flukes, outlined in elegantly bold curving lines, unfurled before us like a flag. Slipping once again out of sight, the calf reappeared by its mother's tail, nuzzling and pushing, perhaps nursing.

Several times, the baby whale swam up to our boat, curious as right whale babies are. Only about 2 weeks old, it was already about 16 feet long, over half the size of our boat. On the port side, I leaned down and watched the calf nudge the bow with its head, which was still smooth and did not yet have clusters of whale lice. Its sides bulged in sleek curves and young dark flesh. It dove under the boat and resurfaced on the starboard. I scrambled across the deck to stare at it again, from barely 2 feet away. I could have reached out and touched it; I was so excited I could have jumped in with it. But I didn't want to touch it. I didn't want to break the magic of the moment. I wanted just to be close to it, feeling it define my space by the immediacy of its presence.

II

The current fashion is to consider the North Atlantic right whale and the South Atlantic right whale as two separate and distinct species, since neither population crosses the equator. If so, the North Atlantic right whale may be the rarest and most endangered whale in the world.

Such a claim is difficult to make. The North Pacific right whale is also extremely rare—for decades, only a handful of sightings of this whale have been recorded along the coasts of Washington, Oregon, and California. Not only is the current estimate of 200 to 250 animals alarmingly low, but it is also a shot in the dark.

The bowhead whale was circumpolar, and it is now given five separate stocks. One of these stocks, the bowheads of the once heavily hunted Greenland and Barents seas, is very near extinction. Two of the other stocks have only about 100 whales left, and another has just a few hundred. The only strong population left, ironically, is the one that received so much media attention because of Eskimo subsistence hunting: the bowheads of the Chukchi and Bering seas. They number between 1000 and 3000. (Before whaling, however, this stock may have contained 18,000 to 36,000 whales.)

According to Randall Reeves, who works at the Arctic Biological Station at Ste-Anne-de-Bellevue, Quebec, and is the coauthor of *The Sierra Club Handbook of Whales and Dolphins*, we don't know just how close to extinction the right whale came. For over a decade, Reeves has studied the hunting records and sightings of the species, and he speculates that the low point came early in the twentieth century with only "a few dozen individuals."

Protected by international agreement since 1937, the right whale in the North Atlantic was known from only a handful of sightings through the 1950s and 1960s, many of those coming from the waters around Cape Cod and Florida. But the reports were so scarce that they seemed to represent the trailing wake of a species headed almost certainly toward extinction.

No whale has gone extinct in modern times. Although many species of whales are cosmopolitan—widely distributed in the world's seas—they form isolated, separate stocks, which are the equivalent of races or nations among humans. One stock—the gray whale of the North Atlantic—has gone extinct, entirely exterminated by whaling. The North Pacific stock of the gray whale, considered extinct on two separate occasions, was compared to the North Atlantic right whale in its precarious status. The protection it received in 1937 was seen as a gesture flung into the teeth of inevitable extinction. Nevertheless, the gray whale along the Pacific coast has recovered: Its current numbers are perhaps 15,000, rivaling the stock's original population.

But if the gray whale on the Pacific coast has recovered so well, why hasn't the North Atlantic right whale? Several recent discoveries of sizable groups of right whales along the Atlantic coast have led to a virtual reinvention of the species. As the enthusiasm grew with increasing knowledge over the last decade, research efforts intensified and led to a slowly evolving picture of the right whale's current status. It is a species we are learning to see and appreciate anew.

On the basis of a catalog of thousands of photographs, Scott Kraus of the New England Aquarium in Boston has identified 230 individual right whales. He estimates there might be anywhere from 300 to 600 in the North Atlantic. Kraus's research is primarily devoted to gathering empirical information on the most crucial, most basic biological questions about an endangered species: How many are there, and where are they?

Underlying the numbers is the simplest, most basic question of all. It is a question not of data, of hard facts gleaned from the field, but of interpretation: Is the North Atlantic right whale making a comeback, or is it still heading toward extinction?

The major breakthrough in studying the right whale, the discoveries that led to the new efforts to understand the species, came in the lower Bay of Fundy. For cetologists, as for the whalers before them, the Bay of Fundy was largely uncharted and unknown, a broad watery wedge separating Maine from Nova Scotia in its lower reaches. A few whalers ventured into the bay, but the historical records show only a scant, casual use of these waters. Mostly, whalers bypassed the bay on their outward-bound journeys, heading for pelagic whale grounds in the Arctic.

Over the last half century, the right whale has been sighted along the Atlantic coast, most notably in Cape Cod Bay. By 1979, right whales were known to be in the Bay of Fundy, with reports coming in from ferry crews and fishermen. On July 10, 1980, flying over the Bay of Fundy while conducting a systematic inventory of cetaceans (whales, dolphins, and porpoises), Scott Kraus and Randall

Reeves found a large number of right whales, their "only real surprise." In that field season, which lasted until October 30, they sighted twenty-six right whales in the area of The Wolves and Grand Manan, including four different cow-calf pairs. The numbers were modest, but for an endangered and unstudied animal, they represented a large concentration and a major discovery.

Realizing that the right whales in the Bay of Fundy were significant, Kraus began a major research effort to learn as much as possible about the species. He and a crew of biologists and volunteers have returned to the Bay of Fundy every summer since 1980.

The second major discovery of right whales in the North Atlantic came from Browns Bank, about 35 miles south of the tip of Nova Scotia. It repeated the pattern of the discovery of the whales in the Bay of Fundy: scant and sketchy historical data preceding a stunning confirmation of large groups of whales from aerial surveys. Flying for the University of Rhode Island and the Cetacean and Turtle Assessment Program (CeTAP), which covered the entire coast in aerial surveys from North Carolina to Nova Scotia, Greg Stone surveyed the area of Browns Bank in 1981. On August 27, he saw forty-six right whales, the largest right-whale sighting in the North Atlantic.

Another important location for right whales was discovered off the southeast coast. In winter, small numbers of right whales— between six and ten pairs of mothers and calves—occur off the coasts of Florida and Georgia.

Scott Kraus has organized research in all three areas on an annual basis—Bay of Fundy, Browns Bank, and the Florida coast. A man with sharp, heronlike features, he told me: "In right whales, nobody knows about them, so we're on the front edge. Not many places you can still do that kind of work. We're still working on the basic questions."

The answers to those questions come from observation in the field and from studying over 16,000 photographs. The study of cetaceans depends upon the identification of individual animals— that is how population size, distribution patterns, and reproductive trends can be determined. Each right whale has a distinctive and unique pattern to the callosities that form on its bonnet and lips. Growing in islands and peninsulas that stay constant for each whale, the callosity pattern is a "fingerprint." Photographs of these callosity patterns are the coin of the realm in right-whale studies: Spending long, dull hours sorting through the photographs, researchers have compiled a catalog of 230 individual right whales.

Using this catalog, Kraus has documented migrations between all three of the major locations so far discovered, and he believes these whales constitute one large population of right whales, not separate groups. Browns Bank seems to be a place for sexually active whales: The whales are active in courtship groups, and Kraus has seen one whale insert his penis into a female, though he carefully points out that this does not confirm that conception took place. In the Bay of Fundy, about half of the sixty or so whales that gather in late summer and fall are cows with 1- or 2-year-old calves. The rest are a miscellany of males and nonreproductive females. These Bay of Fundy whales tend to stay closer to shore. Florida is the only known calving ground for the species, with about ten pairs of mothers and calves seen there in the winter.

A great deal about the species remains unknown. For example, if Florida is the only known wintering ground, with twenty whales, where do the other 200 or more right whales spend the winter? From old whaling records, some cetologists suspect the whales go farther out to sea. Also, females devote tremendous amounts of energy to raising their calves—the right-whale calf may be the fastest-growing mammalian infant in terms of adding body weight. So, after raising their babies, the mothers appear to take a year off, during which they fatten and recoup their strength. Where do the females spend their off year? Finally, Browns Bank is the place for courtship, and if right whales are like other large whales, the gestation period is twelve months—though this is *not* documented for right whales. But if so, why do the right whales give birth off Florida in January? Is the gestation period longer than thought? Or do the whales actually mate in some as yet unknown place in January?

Why haven't right whales returned as forcefully, as prolifically, as gray whales? No one knows for sure yet, but the question is being researched. One fear is that at least part of the reason may be the pollution of the seas. This is the biggest problem now facing marine mammals, and it drives cetologists nearly to despair. The right whale, for example, is living along the near-shore waters of the Atlantic coast in a degraded and assaulted environment. In the Gulf of Maine, where right whales are found, harbor seals have the highest levels of pesticides in their systems of any U.S. mammal. Between 1987 and 1988, 750 dolphins died mysteriously along the Atlantic coast, their snouts and flippers pocked with blisters.

As one marine biologist I spoke with said, pollution has displaced whaling as the largest threat to cetaceans. We don't know exactly what the effects of pesticides and PCBs and heavy-metal contaminants are on right whales. They may suffer from malnutrition or be impaired in their reproductive abilities. But one thing is clear: The pollution cannot be good for them.

III

I have stood on the shores of Cape Cod, watching the sun's thin light fall upon bleached sands and low dunes, looking out on the glaring blueness of the sea, and I have tried to imagine what the early New England settlers saw in these waters. The sunlight doubles back from the sand, glints off the waters, and fills the air with an intense veil of light that is blinding. Though imposing, the dunes and sea are unlike forests and mountains. They are stark and lonely, appealing less to the senses than to the imagination. But it takes a prodigious act of imagination to picture the sea as alive with whales as it once was. The early settlers' journals and reports offer a staggering introduction to a foreign world. The seas are not dead now, but they are less alive than when the first colonists discovered them. When I try to look beyond the light and imagine the lost abundance, I can't do it. I faint under the attempt.

Though researchers are working to determine the size of the right whale's former abundance, numbers only give a plodding and prosaic way to comprehend what must have been some of the richest

waters in the world for whales when the colonists arrived on the
Atlantic coast. One of the pilgrims on the *Mayflower* described the
scene in vivid terms as the ship lay at anchor in Cape Cod Bay, off
Provincetown, in November 1620. In a document now called Mourt's
Relation, the diarist writes of an enormous and unfrightened concentration of what must be right whales: "Great whales of the best kind
for oil and bone come close aboard our ship, and in fair weather
swim and play about us." He goes on:

> And every day we saw whales playing hard by us; of which in that
> place, if we had instruments and means to take them, we might have
> made a very rich return, which to our great grief we wanted. Our
> master and his mate, and others, experienced in fishing, professed we
> might have made three or four thousand pound worth of oil. They
> preferred it before Greenland whale-fishing, and purpose the next
> winter to fish for whale here.

This pilgrim betrays an ominous exuberance in his description, as
if in the moment of discovery the future was defined. If their Cal-
vinism left little room for enjoying the abundance of God's creation,
the austere pilgrims were nevertheless eager to exploit what they
found. With pious eyes fixed on God's face, they were from their
first moments in the New World scheming to pick His bulging back
pocket.

When Richard Mather arrived in Massachusetts in 1635, he
wrote of "multitudes of great whales . . . spewing up water in the
air like the smoke of chimneys and making the sea about them white
and hoary." Yet he became so innured to their abundance that he
betrayed a trace of boredom: The whales had "grown ordinary and
usual to behold." And a memorandum in the British secretary of
state's papers for 1667 reported, "The sea was rich in whales near
Delaware Bay, but . . . they were to be found in greater numbers
about the end of Long Island."

A number of researchers have recently pioneered a renewed
interest in the historical significance of the right whale in North
America: Randall Reeves, on whale records in logs and newspapers;

Selma Barkham, on the Basques in Newfoundland; and Elizabeth Little, on the relations between Indians and the early Yankee whalers. Elizabeth Little works with the Nantucket Historical Association. She divides New England whaling into three categories: drift whaling, along-shore whaling, and pelagic whaling. The typical account of the history of whaling in America moves quickly through the early periods to "the Golden Age of American whaling" in the pelagic period. Little's work is especially important for demonstrating the importance of the right whale in the first two stages of the whale fishery (they were virtually synonymous with the right whale), for demonstrating the rapid depletion of the right whale off the coast of New England, and for untangling the role the Indians played in teaching the early colonists to whale.

If the history of the right whale is largely an indictment of our own depleted inheritance, it is also an act of remembering the past, the prelude to our own remembering of the future.

Most modern histories are inconclusive on the contributions of Indians to the beginnings of American whaling. One tradition has it that the Indians taught the early English colonists how to whale. According to Little, this tradition is a nineteenth-century invention, when writers like Herman Melville displaced the legendary skill of the Indian whalemen backward in time. Though the Indians were famous for their seamanship (Crèvecoeur described them as "fond of the sea and expert mariners"), Little says, "In spite of the tradition, I can find no evidence that Indians of New England routinely killed whales at sea." The only account of Indian whaling during the "Contact Period," when Europeans were exploring the waters and coast of North America, comes from James Rosier, who accompanied George Waymouth in 1605. In cryptic syntax, he describes their custom:

> One especial thing is their manner of killing the whale, which they call powdawe; and will describe his form; how he bloweth up the water; and that he is twelve fathoms long; and that they go in company of their king with a multitude of their boats, and strike him with a bone made in fashion of a harping iron, fastened to a rope, which they

make great and strong of the bark of trees, which they veer out after him: then all their boats come about him, and as he riseth above water, with their arrows they shoot him to death: when they have killed him and dragged him to shore, they call all their chief lords together, and sing a song of joy: and those chief lords, whom they call sagamores, divide the spoil, and give to every man a share, which pieces so distributed, they hang up about their houses for provision: and when they boil them, they blow off the fat, and put their pease, maize, and other pulse, which they eat.

For a number of reasons—location, mention of maize, the unique account of killing by arrows—Little considers this well-known account a muddled, second-hand report.

As Little says, "We have no evidence that Indians harpooned whales at sea off the east coast of colonial America after 1605, until they became involved with English along-shore whaling." She demonstrates that the Indians from Long Island to Nantucket taught the colonists not how to go to sea but how to harvest the whales tossed onto the beaches. The original colonists, after all, were not sailors. The Puritans came expecting to farm and raise sheep. The drift whales became a kind of seashore farming and were harvested more than hunted. To this day, southern New England is the spot in the country with the most strandings of whales. In 1986, one right whale, two humpback whales, about eighty pilot whales, and innumerable dolphins and porpoises inexplicably stranded themselves on the beaches of Cape Cod.

Strandings must have been an even more fertile source of whales in colonial and precolonial times, and Little thinks that drift whales were so numerous that the Indians felt no need to go to sea to kill them. The cetaceans do not drift ashore or strand themselves entirely haphazardly. They are more likely to come ashore at certain beaches than others, and the Indian names of places betray a faint glimmer of the former importance of beached whales in their geography: If the translation is accurate, Siasconset on Nantucket, for example, means "great bones place."

We are likely to feel pathos over the inexplicable strandings of whales and dolphins. We still cannot explain them. These strandings must have appeared in a much different light to the Indians: less a tragedy than a blessing. So significant were these drift whales that they were metamorphosed by the Indians into their mythology. The Algonquin tribes of Long Island, Martha's Vineyard, and Nantucket shared a set of stories about a fabulous culture hero variously called Moshup, Manshop, Moishup, or Maushop.

Moshup was said to have created Nantucket when he knocked the poke from his pipe into the sea. The fog that hangs over the sea south of Cape Cod, obscuring Nantucket, is the smoke rising from his pipe. One of the richest of the Moshup legends portrays him as the Indians' mythic whaleman. The story is quoted from historical sources by William S. Simmons in *The Spirit of New England Tribes*:

> The first Indian who came to the Vineyard, was brought thither with his dog on a cake of ice. When he came to Gay Head, he found a very large man, whose name was Moshup. He had a wife and five children, four sons and one daughter; and lived in the Den. He used to catch whales, and then pluck up trees, and make a fire, and roast them. The coals of the trees, and the bones of the whales, are now to be seen. . . . Moshup went away nobody knows whither. He had no conversation with the Indians, but was kind to them, by sending whales, etc. ashore to them to eat.

It requires a shift in our relationship with nature to understand this story. Since I pay for my groceries, I have the crazy delusion that I have actually earned my salad greens and nacho chips—an attitude inherited from the Puritan work ethic. For the coastal Indians, however, whales were not something they chased, caught, and therefore achieved. Whales were a gift from a vanished hero, a father figure. Drift whales represented the kindness of Moshup, a blessing from their hero, a symbol of their connection with both animal and place. Finding a whale on the beach must have made the world seem like a place of happy chances, and they were the lucky winners.

For the whites, drift whales quickly grew in economic significance. By 1644, within four years of settling in Southampton, townspeople had set forth rules for sharing "such whales as were by hard luck and the kindness of Providence cast up." Deeds on Long Island to important beach areas and the rights to carcasses were always carefully spelled out. The whales, in fact, contributed to a major political controversy between the colonies and England, over a century before Bostonians were protesting the tea tax. When the Dutch handed New Netherland over to the English, Long Island became the special dominion of the Duke of York. Under the "Duke's Laws," whales were declared a "Royal Fish," and a fifteenth of a gallon of oil "shall be received for whales . . . cast upon the Shoare of any Precinct." By 1660, the whales had already become a prime source of wealth, and the fiercely independent islanders resisted this taxation by the Duke's "robber governors." Ignoring the "Perills" from the law, the whalemen contrived to report no whales to their rulers, and thus resisted the taxation.

Because it was more abundant then, and because it frequented the shores so closely, the right whale probably contributed a large percentage of the drift whales on the prime beaches in southern New England. The right whale soon drew the colonists off the shore and into boats. The beaches where the whales had been cast up soon became the places where the along-shore whalers built their lookout masts and launched their boats. Over the next century, the places in New England that had been the best for drift whales became the centers of the great pelagic whaling enterprises. The leading whale ports of the nineteenth century were not the large mercantile or fishing ports of Philadelphia, New York, Boston, or Salem. Rather, they were the ports near the recorded sites of Indian drift whaling, like Nantucket, New Bedford, Sag Harbor, and Southampton. As Elizabeth Little writes, the basis for "the distribution of all three whaling activities [drift, along-shore, and pelagic] was the distribution of right whales near the east coast in colonial times."

The human geography of the eastern coast is also a map of the value of the right whale, demonstrating the significance of the forgotten whale in American history. In a metaphorical as well as an economic way, we built our cities and homes on the back of the right whale.

IV

The move into the sea in pursuit of right whales marked the invention of a distinctly American form of whaling. Following sporadic attempts with little success, the sons and grandsons of the English settlers learned to push through the surf, dare the worst of the winter weather, and challenge a 50-ton leviathan in small boats on rough seas. Goethe writes that the days of becoming are the most invigorating and that the full realization of being always comes with a tinge of sadness for the times of struggle gone by. It must have been an exhilarating time for these early whalers, since they were in the process of inventing a way of life and of learning what they could do.

Part of what makes the enterprise so distinctly American is the relationship between the Indians and the Yankees. Documents now suggest that the English settlers provided the new enterprise of along-shore whaling with the technology and the vision, while the Indians supplied the seamanship, the talent, and the courage. It proved to be a combination that would eventually carry Ahab and Queequeq around the world.

Along-shore whaling was right whaling. As Obed Macy wrote in his 1835 *History of Nantucket*, "The whales hitherto caught near the shores were of the Right Species." He goes on to offer a picturesque story on the origins of along-shore whaling in Nantucket. He wrote, "Some persons were on a high hill, . . . observing the whales spouting and sporting with each other, when one observed, 'there,' pointing to the sea, 'is a green pasture where our children's grandchildren will go for bread.' "

Along-shore whaling was also winter whaling, December to March. Macy described the conditions under which the whalers went to sea: "They sometimes, in pleasant days, during the winter season, ventured off in their boats nearly out of sight of land. It has often been remarked by the aged, that the winters were not so windy and boisterous at that time as at present, though quite as cold; and that it would sometimes continue calm a week or even a fortnight."

By separating figures for along-shore whaling from the composite accounts of early historians, Little concludes that it began in Delaware as early as 1632 and that the industry worked its way up the coast. Charted on a graph over time, the harvests of the various locations are like successive parabolas of plenty and loss. On Long Island, the along-shore whaling began officially in 1650, when John Ogden was granted the first whaling license on record for "free liberty without interruption from the inhabitants of Southampton to kill whales on the south sea." But the colonial whaling industry did not get well under way until 1667, when James Loper, a Dutchman, organized the Indians on the eastern end of Long Island into along-shore whaling companies. The right whales off Long Island were largely exhausted by 1718. In a peak year, eighty whales were killed. On Cape Cod, the boom began after 1680, and it crashed by 1725.

According to Zaccheus Macy, writing in 1792, Nantucket started late: "The whale fishery began at Nantucket in the year 1690. One Ichabod Paddock came from Cape Cod to instruct the people to whale in boats from the shore, and the business lasted pretty good until about 1760, and then the whales gon and pretty much don." The peak year was probably 1726, when eighty-six right whales were taken.

We can get an idea of the kind of riches to be made, and the extent of the destruction of right whales, from some remarkable research on the whaling industry of the Basques on the coast of Labrador. We know the right whale was hunted by the Basques in the Bay of Biscay at least as early as the eleventh century, making the right whale the first whale to be hunted for commercial purposes.

A hardy and mysterious people, the Basques keep popping up unexpectedly in remote corners of the world. In the last decade, Canadian researchers led by Selma Barkham discovered the presence of the Basques in Red Bay, Labrador. In 1977, archaeologists entered the town to find unmistakable evidence of Basque history: in the town's gardens and on the beaches, Spanish red tiles; with just a little digging, flensing knives and cooper's tools; and under water, the biggest prize, a sunken galleon, named the *San Juan*, with 55,000 gallons of right-whale oil in its hold.

The huge yields of oil and baleen from right whales rivaled the more glittering treasures sailing back to Spain in the galleons from the Caribbean. (Environment Canada researchers estimate the hold of the *San Juan* contained about $6 million worth of right-whale oil.) Oil from these whales lit lamps throughout Europe, became tallow in candles, and was an ingredient in products like soap. Plus, the huge plates of baleen in the right whale's capacious mouth ("whalebone," which could reach nearly 9 feet in length), stiffened fashionable women's corsets and put the elasticity into whips.

The Basques had followed the right whale around the Atlantic, whaling from shore in Newfoundland in 1530 and reaching the peak of their success in the 1560s and 1570s. The whole business for the Basques in Newfoundland was probably finished by 1620. There is some debate over how much of the catch was made up of right whales and how much was made up of bowhead whales, a near relative. But with an average catch per boat per year of about twelve whales, one estimate puts the average annual catch at about 300 to 500 whales. The total harvest between 1530 and 1610 is estimated to have been between 25,000 and 40,000 whales.

These seas must have been teeming with whales.

The drogue, or drag or drug, has played a controversial part in the research into early American whaling. Its origins usually attributed to Indians, Elizabeth Little has found its source in the Basques. Before Yankee whalers learned to fasten their boat to the whale with harpoon and line, they used a thick board about 14 inches square to keep the line afloat and tire the whale. According to Little, there is "unequivocal evidence" that the drogue originated with the Basques. In the seal of the city of Fuentearrabia, dating from perhaps 1335, Little has found a square shape attached to a whaling harpoon line flung at a whale.

The question is, how did the Yankees in New England learn to use this drogue. As mentioned, Ichabod Paddock, otherwise unknown, came from Cape Cod to teach the people on Nantucket to whale from shore. James Loper, the innovator in whaling techniques on Long Island, was also asked by entrepreneurs on Nantucket to aid them in establishing whaling companies. Little says that if we wish to speculate, perhaps Ichabod Paddock or James Loper introduced an ancient Basque technique to the Indian and Yankee whale crews. It was both intelligent and successful. As she writes, "Successful colonial American whaling appears to have begun only after the American Indians learned an archaic European method of catching whales."

If Indians did not supply the equipment for "fixeing out whaling," they indisputably taught seamanship to the New Englanders. In energetic prose, William Wood of Salem conveyed his astonishment at the skill of the Indians at sea in 1635, calling the Indians' canoes "these cockling fly-boats, wherein an Englishman can scarce sit without a fearfull tottering." According to Crèvecoeur, the Indians of Nantucket dominated colonial American whaling. In a whaling boat of six men, five would be Indians and one would be white, and the white one would be the leader.

On the traditional drift-whale beaches, the whalers erected a mast or lookout, from which they watched for the spouts of whales. Huts or whale houses crouched on the shore around the mast—thatched wigwams on Long Island; small, single-story, boarded houses on Nantucket. At the sight of a whale, the lookout cried, "*Awaite Pawana*, here is a whale."

With what Crèvecoeur calls "astonishing velocity," the men tried to pull within 15 feet of the whale—close enough for the poised harpooner to try for the thrust of death. With one leg braced in a notch in the bow, he aimed for the prime spot just behind the whale's head. In pitching seas, often in miserable winter weather, in a 20-foot boat beside a 50-foot whale, these men faced terrifying dangers. The small boat gave the men the advantage of mobility; nevertheless, to quote Crèvecoeur, "Sometimes in the immediate impulse of rage, she will attack the boat and demolish it with one stroke of her tail. In an instant the frail vehicle disappears and the assailants are immersed in the dreadful element." A man who fell overboard in the winter seas stood little chance of survival if he was not immediately hauled from the waters.

The final struggle could last for hours as the whale fled or fought for life, slowly wearying from the assault. The boats closed in as the whale, pulling the drogue, tired. The men used sharp lances for the death stabs, until, in the last paroxysm, the whale spouted a crimson blow, her breath thick with blood—the sure sign of death.

Sometime before 1782, New England whalers learned to fasten the harpoon line to the boat. This technique had also been developed by the Basques, and it was used by English and Dutch whalers near Greenland. Thus began the "Nantucket sleigh ride": A captain had grown tired of losing harpooned whales and tried to convince his men to use a new method, just told to them, "by boat and the line." According to Thomas Beale, who in 1839 published *The Natural History of the Sperm Whale*, the idea "seemed monstrous; the mere thought of having the boat they were in attached to an infuriated leviathan by a strong rope struck terror among the whole crew. . . . Others more daring undertook the trial soon afterward, in which they frequently came off victorious, so that the new method was established among them, and has since been much improved."

The along-shore whalers were too successful. Early in the 1700s, complaints began to surface that the whales were no longer in their usual haunts. But emboldened by their success with right whales, and hearing that the sperm whale, once thought rare, was abundant far offshore, the men on Nantucket began by 1718 to go after it. The sperm whale soon displaced the right whale as the preferred creature of the hunt.

Right whales continued to be hunted from the shore through the nineteenth century: About ninety were taken from Long Island in the last half of the century. The last one was killed off Long Island in 1918. But for the most part, whalers had turned their attention to other species and other seas. With an awful inevitability, one species after another fell under the onslaught, and even in the mid-1900s Herman Melville wondered whether "Leviathan can long endure so wide a chase, and so remorseless a havoc; whether he must not at last be exterminated from the waters."

As the waters around the world were depleted, and as the preferred species went into their predicted declines, whalers turned to new hunting grounds and new species. Advances in technology (the heavy cannon harpoon gun, faster ships, and floating factories) made it both efficient and profitable to take species that had formerly been too small or too elusive to be considered worth the pursuit. This particularly meant hunting rorquals—blue, humpback, minke, fin, and sei whales. They were the untapped reserve of the "wrong" baleen whales of earlier times.

Protection for the whales came late, and it is a sad case of trying to close the barn door after the horse has bolted. The International Whaling Commission (IWC), founded in 1946, has been plagued by controversy and ineffectualness, but it has gradually extended its protection over the large rorquals, right whale, gray whale, bowhead whale, and sperm whale. In 1983, the IWC made the historic decision to impose an indefinite moratorium on all whaling. Sadly, whaling has probably been discontinued not out of concern for the whales but because the enterprise has lost its profitability. There are still indigenous subsistence whale hunts—notably by Inuits for bowheads and Faroe Islanders for pilot whales—though this pressure on the populations is not now a major concern for most environmentalists. Only one member country of the IWC now refuses to honor the moratorium—Japan.

Though protection has helped the whales, most species are appallingly low fractions of what they once were. Even if no species has gone extinct, several of the stocks or regional populations have vanished or are in imminent danger of doing so:

Right whales: Precarious throughout their range, and on the verge of extinction in the North Pacific and North Atlantic.

Blue whales: A few hundred in the North Atlantic; of the 5000 once in the North Pacific, perhaps 1500 survive; of the 200,000 in the southern hemisphere in the nineteenth century, a recent scientific census saw only 453 animals, and estimates only 1200–1500 blues remain.

Humpback whales: Of 15,000 in the North Pacific before the onset of mechanized whaling, less than 1000 survive; the western North Atlantic has about 2000, though the eastern North Atlantic is in poor shape; in the southern hemisphere, of the 100,000 humpbacks in the nineteenth century, about 2500 are all that remain.

Fin whales: Severely reduced throughout the seas, perhaps by one-half to two-thirds.

Sei whales: Heavily exploited, suffering major declines in all areas.

Minke whales: Heavily exploited, largely for their meat, though they were once considered too small for whalers to waste time and effort on.

We have inherited, and created, impoverished seas. The spouting is more sporadic, gone totally from some areas of the oceans, and the whales that remain are mere relics of the swarms our ancestors wrote of with such awe. Whaling, now a small industry in terms of former glories and actual contributions to national economies, is conducted only in defiance of world opinion. Still, one nation refuses to abandon a business that has become an ethical and economic anachronism—the sad devastation for dollars that began with the right whale.

V

Lately, my dreams have been filled with shorelines. I walk or drive or even fly along the edge of the water, drawn to enter, but afraid. In one dream, I was returning from a long trip, but the airplane could only land at the more unusual Boeing Field in Seattle, not Seattle-Tacoma International Airport, and the 747 taxied absurdly along the rocky shores of Puget Sound on the way to my house. When I got home, my ex-wife was waiting for me, looming in the doorway, holding a monstrous dead salmon up by the tail.

If we were honest, many of us would have to admit that leaving is easier than coming home. I'd rather keep running, keep moving, keep my hands on the wheel. Yet there's something else, too, something closer to home that I can't quite leave behind. These shorelines, with their ugly fish-facts looming in the doorway. The right whale is a national fish-fact—ugly, battered, forgotten as we moved on. I'm torn between these two impulses: to go, and to return. Always restless.

Who doesn't have ugly fish-facts waiting for them at home? Or in their hearts? Yet I know few people willing to face themselves. Instead, we seem bent on some mad national hypocrisy. The Faustian dream that drove Captain Ahab isn't dead, it has just grown sillier in the way we live it out. Ahab at least went mad in his hate for the whale that maimed his body, that stole his leg. He chased it with a "frantic morbidness" as the incarnation of "not only his bodily woes, but all his intellectual and spiritual exasperations." There is something heroic in his excesses. Thus, Melville makes it clear that Ahab's hatred was a monomaniacal projection of all evil, of an inner loathing, "visibly personified, and made practically assailable in Moby Dick."

But us? We can measure our pretensions and ambitions with a "frequent flyer plan." Everyone pretends to be going somewhere, but all this traveling is just a debased version of the old humanist ideal of self-creation: defining ourselves not by where we are but by where we think we're going, not by the present but by the future, not by what we are but by what we prefer to believe about ourselves. Everyone's bursting with the latest hype from some minister of pop psychology or positive thinking.

Picture this: I get on an airplane. Everyone's got his or her wish: We're all flying executive class, all going to the same business meeting in Atlanta, all reading the same book—*Celebrate Your Self.*

The very thought of so much narrowness and denial, of a culture so obsessed with images and money, makes me claustrophobic. Americans are so competent and so noble in their aspirations, so boring and so upwardly mobile. What these bourgeois platitudes of self-creation boil down to, in my experience, is that we're all free to choose our own hairstyles.

We're all free to choose our favorite illusions.

When I traveled to Cape Cod, early one September, to see right whales there, it was with a sense of disappointment. I live enough in the grip of the fantasy of the heroic, I am so full of the compulsion for the remote, that seeing whales out of Provincetown seemed too tame. I would have much preferred to see right whales with researchers up in the Bay of Fundy, off the coast of Maine, or, better yet, out at sea, on Browns Bank. But at the time, that was impossible, and I was reduced to riding on Dolphin Whalewatch boats with the late-summer tourists. It took me days to begin to get in step with my own experience, to quit trying to relate to the whales in some heroic way, to stop trying to imagine myself as Ishmael plunging blindly "like fate into the lone Atlantic."

On the fingertips of Cape Cod's curling arm, Provincetown in the summer is a transient colony of artists and tourists, gay men and women, whale researchers and fishermen. In the dog days of late summer, its streets were full of the seething energy of a place out of the mainstream. Stormy Mayo lives here year-round, in a clapboard house overlooking the inner harbor. He directs the Center for Coastal Studies (CCS). It's a private research organization, and Stormy is an energetic hustler. His mind is nimble, and he parodies his own loquacity as he talks. He likes bizarre ideas, and talk for him seems generative, a way of inventing possibilities. In the early 1980s, after meeting Elizabeth Little on one of the tour boats that the CCS uses for some of its research, Stormy realized that right whales might be wintering in Cape Cod, as they did in colonial times. In 1984 he began looking, and he calls the right whales he found "the single biggest contribution that the Center for Coastal Studies has made to science."

I had one particularly memorable evening with Stormy, full of beautiful light and a feeling for the whales. Stormy led a group of tourists out to the right whales, exciting them about seeing "the most special animals on earth." Wearing a cranberry plaid shirt, a double-billed whaler's cap, and sunglasses, he seemed dressed for the part. He raced around the boat, took pictures, gasped when whales surfaced nearby. People crowded around him, eager to ask questions, to tell him what *they* knew, feeding off his contagious enthusiasm.

It was the mystery and mystique of the whales that fueled that enthusiasm. He said, "We catch a glimpse of the whale. But what do we see? Maybe ten minutes, all told, out of what may be a forty-year life of the whale, in thousands of miles of ocean. Maybe all we see is atypical. A whale grabbing a breath. But the whale's under water most of the time, and who knows what goes on under water. So we can't really be scientists."

We all oohed and aahed when we saw whales. Fin whales swam near the boat, their dark, sleek bodies rolling like wheels just breaking the surface of the water, their dorsal fins like small hooks on their backs. We watched them lunge-feed, their mouths agape. One side of their jaws was white, shining under water as the whales crashed along the surface like open-ended barges. Sei whales had come close to shore that year too, and we saw them—a rare event, since they usually keep farther out to sea. They look much like fin whales, except they lack the white on their jaws, and their dorsal fins are more acutely angled.

But without question, the right whales stirred the greatest passions, because of their size. When a year-old calf came right up to the stern, everyone on board crowded close to see it. I could feel the whales give energy to everyone on the boat. Their mere presence was a magnetism.

There must be deep psychic roots to this power the whales exercised over us. It was more than just the spectacle of these massive creatures. The whales evoked a vitality that can't be intellectualized, though I listened to lots of people on the boat who were trying to confine the whales we saw to categories and information. Our enthusiasm had to come from some primal sources, because it came out as such pure vitality.

I think the energy came straight from the body. These huge phallic beasts swimming through the wet sea. These big gaping mouths. These massive bulging bodies. There was no resisting them. The enormous mass of flesh. The power of the body.

Nature just is, I suppose, but what we make of it depends on mood and culture.

The biblical story of Jonah is an extended meditation on the physical features of the whale, especially the mouth and belly—the "great fish" which came to "swallow Jonah." Thrown overboard, Jonah laments his sinking into the depths, the *de profundis* of so much Christian literature: "The waters closed me about, even to the soul; the depth closed me round about." And Jonah finally "cried out of the belly of hell."

The whale and its mouth and belly became part of the Christian iconography of hell: It symbolized an appropriate punishment for being swallowed by the world. Pieter Brueghel the Elder uses the whale's open mouth as the gate of hell, and the belly represents suffering for the outcast souls.

Even as late as the nineteenth century, Rudyard Kipling focuses on the mouth of the whale in "How the Whale Got His Throat." A clever sailor was swallowed by the whale, and while inside, stuck his demolished raft into the whale's throat. This subsequently prevented the whale from eating anything except "very, very small fish; and that is the reason why whales now-adays never eat men or boys or little girls."

And, of course, there is *Pinocchio*. The little boy disobeys his father, skips school, and follows his own willful desires. Looking for him, his father is swallowed and nearly lost in the belly of the whale.

A profound oral nightmare, a primal Freudian terror—being swallowed up. The life of desire and the body, of the sea and nature, is deeply dangerous. If Eros is one of the main attractions of nature, the opposite side is the fear of death, Thanatos. We will be swallowed by our own appetite for life.

But this is viewing the body from the outside, fearing the flesh that will ultimately betray each of us in death. It is a moralism that hopes to control the body, and control nature, out of fear. It tries to lift us out of nature and a part of ourselves that loves the world too much.

But the body is as well a release, an intoxication with all the loveliness of desire. Cetaceans represent the erotic body as well as the threatening body. The ancient Greeks loved the association between cetaceans, the flesh, and desire. In the Musea Nazionale, in Naples, there is a stunning statue of an elegant youth with lovely long hair, entwined with a large dolphin, not being swallowed by it but embracing the creature. It is Eros riding a dolphin. The dolphin's tale wraps sinuously, powerfully, around the upthrust legs of Eros.

The naked youth with dolphins is a common classical motif. You cannot travel in Italy without finding statues—ancient or

Renaissance—of naked boys playing with dolphins. *Faun on a Dolphin*, in the Villa Borghese in Rome is ancient, a statue of a slender youth with curled hair riding a dolphin's back, the boy's torso twisted with a hint of power.

Venus also is associated with cetaceans. An Albrecht Dürer woodcut depicts her on two scaly, fierce whales. She is called "Venus Marina," and Dürer portrays her as clothed, rather chaste, but with her robes and hair flying; she is holding on to full, billowing sails.

What is Venus's birth? Rising from foamy seas, escorted by leaping dolphins, Aphrodite, or Venus, is herself the breaching whale, pure sexual energy rising up and breaking the surface. Sandro Botticelli's famous *Birth of Venus* has no cetaceans in it, but his sensual and modest Venus is carried on the sea in a seashell. Other painters make the connection between whale, sea, and Venus more explicit. Jean-Auguste Ingres, for example, offers a gorgeous Venus, her long hair decked with pearls, arms draped over her head, shoulders open, body supple in its full curves. She stands amid the frothing sea, *putti* at her feet, accompanied by lovely, happy dolphins.

Dionysus, the god of intoxicated desire, of the body irrational, is associated with the sea and with small whales. Oppian writes in *Halieutica*: "But diviner than the Dolphin is nothing yet created; for indeed they were aforetime men and lived in cities with mortals, but by the devising of Dionysus they exchanged the land for the sea and put on the form of fishes. . . . What a marvel shalt thou contemplate in thy heart and what sweet delight, when on a voyage, watching when the wind is fair and the sea is calm, thou shalt see the beautiful herds of Dolphins, the desire of the sea; the young go before in a troop like youths unwed, even as if they were going through the changing circle of a mazy dance."

In the Staatliche Antikensammlungen und Glyptotek, a museum in Munich, is the most beautiful bowl I have ever seen. In the sixth century B.C., Exekias produced one of his best-known masterpieces, showing a black-figured Dionysus reclining on his left arm in a boat on a red sea. The slender mast supports twining vines laden with ripe grapes, and the sea swims with leaping dolphins. Transformed from Tyrrhenian pirates, the small whales jump about him in libidinous play.

This tradition reaches ecstatic levels in D. H. Lawrence. In "They Say the Sea Is Loveless," he writes of Dionysus and whales:

But from the sea
the dolphins leap round Dionysos' ship
whose masts have purple vines
and up they come with the purple dark of rainbows
and flip! they go! with the nose-dive of sheer delight;
and the sea is making love to Dionysos
in the bouncing of these small and happy whales.

In his *Studies in Classic American Literature*, Lawrence writes of Melville's white whale: "He is the deepest blood being of the white race; he is our deepest blood nature. And he is hunted, hunted by the maniacal fanaticism of our white mental consciousness . . . Hot-blooded sea-born Moby Dick. Hunted by monomaniacs of the idea."

In Lawrence, the whale as body rises beyond eroticism. He celebrates the mysteries of a new kind of consciousness—the transformation of the white mind through the mysteries of body consciousness, the body as the vehicle of the transformation of psyche. We are pirates transformed by Dionysus into leaping whales. More than mysticism, this is wild, urgent prophesy—singing of beginnings and ends, of the creation of new lives out of the fierce burning of desire. The immense bodies of the whales are prophesies of a new age, full of the passion for life, not for money. That is the possibility opened up by "Whales Weep Not!"

They say the sea is cold, but the sea contains
the hottest blood of all, and the wildest, the most urgent.

All the whales in the wider deeps, hot are they, as they urge
on and on, and dive beneath the icebergs.
The right whales, the sperm-whales, the hammer-heads, the killers
there they blow, there they blow, hot wild white breath out of the
sea!

And they rock, and they rock, through the sensual ageless ages
on the depths of the seven seas,
and through the salt they reel with drunk delight

and in the tropics tremble they with love
and roll with massive, strong desire, like gods.
Then the great bull lies up against his bride
in the blue deep bed of the sea,
as mountain pressing on mountain, in the zest of life:
and out of the inward roaring of the inner red ocean of whale blood
the long tip reaches strong, intense, like the maelstrom-tip, and comes
*   to rest*
in the clasp and the soft, wild clutch of a she-whale's fathomless body.

And over the bridge of the whale's strong phallus, linking the wonder
*   of whales*
the burning archangels under the sea keep passing, back and forth,
keep passing, archangels of bliss
from him to her, from her to him, great Cherubim
that wait on whales in mid-ocean, suspended in the waves of the sea
great heaven of whales in the waters, old hierarchies.

And enormous mother whales lie dreaming suckling their whale-
*   tender young*
and dreaming with strange whale eyes wide open in the waters of the
*   beginning and the end.*

And bull-whales gather their women and whale-calves in a ring
when danger threatens, on the surface of the ceaseless flood
and range themselves like great fierce Seraphim facing the threat
encircling their huddled monsters of love.
And all this happiness in the sea, in the salt
where God is also love, but without words:
And Aphrodite is the wife of whales
most happy, happy she!

and Venus among the fishes skips and is a she-dolphin
she is the gay, delighted porpoise sporting with love and the sea
she is the female tunny-fish, round and happy among the males
and dense with happy blood, dark rainbow bliss in the sea.

Next to the ponderous whales, next to Lawrence's poetry, the modern cult of the body is petty and trivial. The love of the body has been displaced into images: We love pictures of the body. Not the body with its flaws and its flesh, but the emaciated body, fed and starved at the same time. The women with anorexia nervosa are simply the most conspicuous victims of a new asceticism, beating the body like the monks of old in more modern forms of punishment and discipline.

As the Dolphin Whalewatch boat headed back to port along the coast of Cape Cod, the air thickened in the cool evening to a magical mistiness. The light danced off the wind ripples of the ocean, fractured into a thousand hypnotic suns, and filled the air with golds and silvers and flecks of fuchsia. The last right whale we saw lay on the surface between the boat and the low outline of Cape Cod. Gulls screamed and circled above her resting bulk. Lifting her head, she lunged forward, shooting waves and bubbles about her slightly opened mouth, like the yawning of desire, and made a strange, plaintive, prophetic call, a mooing sound that seemed to catch and hang in the rich colors of the evening air.

She went under, a shallow dive, rolling forward to get ready for a deeper sounding. She was about to plunge into deeps we could never know, and this was preparation, gaining momentum and rhythm. We all knew it, and braced ourselves for the pure pleasure of her dive.

She reappeared, head high and covered with callosities and white patches, mouth closed in a huge smirk. This time her long slow body lifted high in the water, arching behind her head like firm and supple rubber, bending powerfully into the sea. At the last moment, her flukes rose straight out of the water, for one vibrating and pendulent moment, erect and stiffening, sea water dripping from her wide and waving tail.

I watched her and thought of the times I have stared at my own hands, moving them and marveling at them and at how unknowable my own body is. It is what's given to me, and it is the threshold to bottomless depths and shadows. And if we are all responsible for finding our meanings in nature, knowing as we do so that every moment entails its own projection, I know I found one of mine in the peaked flukes of this whale. If the right whale issues any summons at all to a new consciousness, it is a plunge into the depths.

The whale slid into the sea, slick and silky. Under water, she gave another big thrust with her tail, leaving a round swirling print on the surface, a vitrescent mantle in the sea of gold.

VI

The body connects us to nature and gives us life, but it is also external to us, always limited and known only by its surfaces. The body is an empirical datum of our lives, but our way of understanding the body renders it no more real than any of the other illusions that modern philosophy has found so easy to destroy.

Michel Foucault argues in *The Order of Things* that the modern emphasis on empirical knowledge is largely a creation of the late eighteenth and early nineteenth centuries. René Descartes is often considered the enemy of an enlightened understanding of animals. Descartes is the seventeenth-century philosopher who bequeathed to the modern world the separation of mind and body, humans and nature, subject and desire. Animals for Descartes were relegated to merely physical existence, and, lacking mind, were thought to operate as machines, reducible to a world of cause and effect.

Foucault argues, instead, that a new *episteme*, or way of knowing the world, emerged in the empiricism of the nineteenth century, displacing Cartesian rationalism and making possible the disciplines we call biology, philology, and economics. "We tend to imagine," he writes, "that if these new domains were defined during the last century, it was simply that a slight increase in the objectivity of knowledge, in the precision of observation, in the rigor of our reasoning, in the organization of scientific research and information—that all this, with the aid of a few fortunate discoveries, themselves helped by a little good luck or genius, enabled us to emerge from a prehistoric age in which knowledge was still stammering out . . . the classifications of Linneaus."

Rather, Foucault claims that a radical shift in consciousness created the modern view of nature. He claims that the rise of empiricism corresponds to the rise of history as a primary discipline.

We have anchored life in time, and thereby affirm its reality. This is what gives creatures, for example, their particular mode of being in our knowledge. This way of knowing creatures has given us modern biology and our way of recognizing a creature's existence.

Note how this applies to our knowledge of right whales, of ferrets, or, in the next chapter, of ivory-billed woodpeckers. Unless we can see them, unless we can count them, unless we can put numbers to their population and grant them this empirical existence, they are somehow not real to us. For example, the use of empirical methods with the right whale—the combination of observation and photographs—did more than simply confirm that right whales survived in the North Atlantic. It was our way of granting existence to the whales, of validating their existence. Our data, in some basic way, made them alive for us. Until we had that confirmation, the right whale for us did not exist even to protect. How else can we understand our strange discomfort at not knowing if a creature exists or is extinct, like the ferret or the ivory-billed woodpecker? How else can we account for the American reluctance to set aside habitat, such as national parks in Alaska, for animals that we may never see? That is why we remain vulnerable to the old argument against protecting spaces for animals: that only a few people will get to see these creatures, and what good is that.

Foucault also argues that understanding animals as empirical creatures—a "constitution of living historicity"—also left a void in our apprehension of them. The creatures became at once visible and invisible to us. Some essence in the creature's life became inaccessible—its "organic structure," its life, was lost in the recesses of the past time where it was generated. The visible world has become a superficial glitter over an abyss. Now, Foucault writes, we must also "direct our search towards that peak, that necessary but always inaccessible point, which drives down, beyond our gaze, towards the very heart of things." Animals offer themselves to us now from a buried depth: "It is from the depths of the force that brought them into being and that remains in them, as though motionless yet still quivering, that things—in fragments, outlines, pieces, shards—offer themselves, though very partially, to representation."

Nor are humans exempt from this paradox. We arose in the "space hollowed out by living beings." We make the creatures exist in a certain way that doubles back on us and confirms that we exist like them—creatures, bodies living in time, studied but eluding ourselves always in the moment of recognition. We are the animals' double—the "enslaved sovereign and observed spectator," as Foucault puts it.

For Foucault, the exalted image of man is dying in its own contradictions. He heralds the end of humanism, with its simultaneous domination of nature and alienation from it, sustained only by the power it confers to control and exploit. As the older image of humans dies, Foucault claims, a void is created, and modern thought takes its place in that void. Modern philosophy, he says, must offer a subversive sneer and laugh from that void. Working on the edge of what can be conceived by our own forms of thought, Foucault is part prophet for a new consciousness—new ways of conceiving both ourselves and the world of nature. What this new consciousness may be, Foucault does not say. Nor is he able to say in advance what the new road into the future will be. But he suggests we are living in an exhausted culture, and he points us toward the abyss under our vision, and toward the Other.

Perhaps Foucault is right. The will to power has given us immense control over nature, but it has also given us a planet increasingly out of control. Endangered species are both a problem in themselves and one example of the way our effects outrun our designs. The pollution of the seas, the devastation of the rain forests, the destruction of the upper atmosphere, the threat of nuclear holocaust—perhaps we can contain these problems by our high-tech responses. But I doubt it. We've known about all these for years, yet they keep getting worse. As Gary Snyder says in "Mother Earth: Her Whales":

> *How can the head-heavy power-hungry politic scientist*
> *Government two-world Capitalist-Imperialist*
> *Third-world Communist paper-shuffling male non-farmer jet-set*
>   *bureaucrats*
> *Speak for the Green leaf? Speak for the soil?*

Where might the dream of self-transformation lead? One of the places it took me was to the southern tip of Nova Scotia, looking for right whales on Browns Bank, 35 miles into the Atlantic Ocean.

I joined Scott Kraus and his crew of biologists and volunteers in their late-summer, early-fall research on the right whales in the Bay of Fundy and on Browns Bank. From July to October, he bases his fieldwork out of Lubec, Maine, the northeastern-most city in the United States. The crew is made up largely of volunteers, semino-madic people really, working for nearly nothing just to live near the whales. Some were just out of college, looking for something to do that mattered to them, that was more than making money. Others worked flexible jobs so that they could follow the whales up and down the coast during the year—winter in Florida, summer in Maine.

One man, named Brian Hoover, was planning to go to graduate school soon as a student of marine biology, and he had plans to look for the probably extinct monk seal in the Caribbean. The next year, his sense of adventure led him to climb Mt. McKinley in Alaska, where he was killed; his body was never recovered.

Lubec is a small fishing town past its prime, its modest houses sprawled over a small hill above Passamaquoddy Bay. Some of the volunteers lived on the second floor of the brick-red cannery by the water. Others of us lived in one of the grand old houses higher on the hill, dilapidated and dirty, plaster falling at night from the ceilings while we slept. Though it was rundown, it gave us magnificent views of the harbor.

From Lubec, seven of us formed a research team on a rented yacht with (of all things) teak paneling and a tape deck. We dared a wild passage on heavy seas across the Bay of Fundy, then skirted the inner coast of Nova Scotia from Briar Island to Cape Sable Island, where we made port in tiny Donald Head. Just a fishing village, it had a new harbor crowded with beautiful, broad-beamed, and dis-tinctive fishing boats—Cape Sable Islanders, painted in vivid pastels of blue and red and yellow, all of them using as well their charac-teristic shade of chlorine green.

Bad weather, combined with a plague of boat problems kept us in port for a whole week; our silly yacht, rich in comparison to the

Islanders and always broken down, was stupidly conspicuous. We drew two visits from customs officials, who were convinced we were drug runners.

One afternoon we forced ourselves out to sea under iffy circumstances: The sky was low, and the fog had not fully dissipated. But we wanted to get out among the whales. About 35 miles south of Cape Sable Island, between Browns and Baccaro banks, we ran into a group of at least eighteen right whales—it was hard to be sure of numbers, because the whales came at us so fast from every direction, surfacing and diving. Browns Banks is a courtship area for summer right whales, and they were active and busy.

I scooted down onto the foredeck, clinging to railings, and braced myself on the bowsprit. Always, it's a dislocating experience for me to get right down close to the ocean's surface, to feel small above the hundreds of feet of ocean below me, with nothing but miles of empty, steely ocean visible in all directions. Yawing and pitching with the swells, I dipped with the bowsprit to within inches of the sea.

At that instant, barely 10 feet from me, two huge whales broke the surface with an explosive whoosh, their characteristic V-shaped blows rising like the steam from vents on city streets, their backs flat and black as asphalt sidewalks. Fat-cheeked under the inward curves of their scalloped lips, they had mouths and snouts of startling size, maybe 15 feet long. We were so close, and the moment was so suddenly intimate, that I could hear the water dripping from their jutting lower jaws as it splashed into the vastness of the sea.

It was a jarring contrast: two massive and moving bodies, right there before me, suddenly defining the depth of field for all the empty wilderness of water around us.

They wallowed on the surface. One of the whales lifted a flipper, flat and splayed, and flailed it against the waves, and I could hear it slapping. Dropping out of sight, the whale rolled onto its back, the underside of its belly bursting out of the water, flippers at angles to each side throwing white spray. Unlike the rest of its black body, the belly was a blaze of white, as if it had been splashed with paint. Even in this dark sea, the water washed off the white blaze in ice-blue streams, clear and bright cascades.

The second whale submerged and then rose under the first. They turned belly to belly. Slow but forceful, their big bodies banged indolently, and a white froth stirred the waters around them.

Then one of them swam right up to me in an enchanted moment in the midst of the sea, intensely private and peaceful. It lifted its chin out of the water, and on the underside of its cyamid-covered head, a white blaze lit its chin. It hung there, poised, for a moment. Then it slipped backward into the water and expelled a blow. The vapor of its breath trailed just over the gray waters and drifted into my face and onto my lips.

I licked my lips and tasted the salt. The whales submerged and swam off. In the calm of that encounter, the simple peace in the midst of such huge stretches and such awful deeps of sea, there was truth for me. In fragile moments in the midst of the void, in the fleeting encounters with another living being, in the delicate feel of the breath of another on my face, I can feel my heart open up and the earth become less desolate.

I shuffled my way to the bridge, where the view of the whales was strikingly different. From the bridge, intimacy gave way to energy and power, whales visible in nearly every direction. Flukes wavered above the swells, peaks slipping beneath the peaked waves. Whales rolled over each other, more whales swam into view, whales bellied up to the surface, whales lay motionless in the lapping swells, and whales thrashed a frothing sea. Twisting, touching, stroking with flippers—the scene was too dynamic and active to comprehend fully in any one look.

Even out on Browns Bank, well into the sea, these huge whales seemed fragile and in jeopardy. These courtship grounds are right next to major shipping lanes in the Atlantic. Barely recovered from near extinction, many of the whales were banged and brutalized. We could identify many of them by scars from their collisions with boats. One of them, which the biologists had appropriately named Creases, had a gash of white scar tissue down its back; more than a meter long, the cut had probably been caused by a propeller. In recent years, three whales known to researchers had been found dead from collisions with ships.

As we were photographing and recording, the latent power of

the whales broke into a thrust of pure energy. About 3 miles off
starboard, through the gray sea-air, a right whale rammed up out of
the sea, breaching in an extravagant release of a body out of water,
physical life bursting from the elements that had contained it.
Though, swimming in the water, the right whale is sluggish and
slow, in the instant of the breach this creature seemed transformed
by its own explosion of energy. Straining into the jump, its massive
body was bent and unavoidably phallic, and water fell from its sides
like clouds.

We headed straight for that whale, forgetting data, forgetting
all the whales closer by.

Defying water and air and probability, the whale—all 45 tons
of it—breached several times. It launched almost completely out of
the water, twisting in muscled torsion, quivering on the axis of its
tail, and then crashed back to the sea on its side, white sheets of
water avalanching before it like snow off a mountain.

The breaches were like a storm breaking open and disturbing
the surface of life, like the primitive force that every now and then,
for all of us, heaves up out of its confines, insists on its own expression, and then falls back into itself. They were like life itself, flashing
for one shining instant above the gulf in a superb display of will and
strength.

By the time we got there, the whale had quit breaching. All its
flinging into the air, all the energy of its body expressed in those
fleeting and precarious moments above the yawning sea—they were
gone, leaving only a vibration in the air.

The breath on my lips. The breach from the sea. I need the
friction of these opposites. They give me energy, light me up. Moments of stillness and communion, acts of passion and flurry: breaking in and breaking out. I keep veering between the passsion and
the peace, the future very much in doubt, and I don't understand.

NINE

RUMORS OF EXISTENCE

I

We are living in an age of loss—more loss, even, than occurred in the Pleistocene. I want only to see the truth about life in our times. I'm not interested in making sure I feel good about myself, and I don't think that focusing on loss is simply negative thinking. It is honesty—about life and about ourselves. Extinction has become a part of the meaning of our lives. It's happening around us. If you just look, you can literally see it happening. You're a witness to death on a scale unknown before in history and prehistory. There is really no debate, except over the extent of the catastrophe: Will the rate be 40,000 species per year in the year 2000, or will it be half that? David Ehrenfeld, professor of biology at Rutgers University, has argued that the extinction rate is now 1000 times greater than it was in the Pleistocene epoch. Daniel Simberloff, a professor of biological science at Florida State University, has estimated that by the time two-thirds of the tropical forests are cut, we will have lost 625,000 species. Other estimates are higher.

We might as well face it. We live in a new age: the age of mass extinction.

Come closer to home. According to Paul Opler, a biologist with the U.S. Fish and Wildlife Service, in one bad period of the Pleis-tocene, over a 3000-year stretch, North America lost about fifty species of mammals and forty species of birds—about three species per hundred years. Since the Puritans landed at Plymouth Rock in 1620, however, over 500 species and subspecies of native animals and plants have passed into extinction.

If the mass extinctions culminate as projected, human popula-tions will crash too. We are a species engendering its own collapse.

Wall-to-wall loop-pile carpeting and the best dinner china, VCRs and compact-disc players, large houses and fast burgers—all the accoutrements of bourgeois comfort are cushions to keep us from this frightening truth: We make other creatures pay the price for our prosperity. Lulled to complacency by platitudes and self-deceptions, and just plain weary of the burden of consciousness, we try to forget that those very comforts are purchased by the deaths of right whales and red-bellied turtles and silverspot butterflies. What we are watch-ing, in our times, is the displacement of life, through extinction, into comfort and security.

So I find myself brooding on the meaning of life in an age of loss. Endangered and extinct species are symbols of the darker side of modern culture, in both our external and internal landscapes. Endangered animals have repeatedly taken me to the place where outer swirls back toward inner, where perception melts into feeling. They represent the kind of meager, minimal life we have come to tolerate and expect in the world. The attempt to restore these lost creatures is nothing short of the quest for what I am always looking to find in animals: the reaffirmation of life at deeper levels than I normally experience.

But I am left with elegies in an impoverished world. In "The Oven Bird," Robert Frost poses a question about poetry in our time, but it also applies to the loss we are witnessing in nature and wildlife. The bird is singing at the end of summer, and Frost interprets his song:

> He says the early petal-fall is past,
> When pear and cherry bloom went down in showers

On sunny days a moment overcast;
And comes that other fall we name the fall.
He says the highway dust is over all.
The bird would cease and be as other birds
But that he knows in singing not to sing.
The question that he frames in all but words
Is what to make of a diminished thing.

We live in a larger fall, the fall of nature and the fall of the spirit, with highway dust covering us in a thin film of forgetfulness. The ovenbird's question is our question: How do we live with less?

This raises major political as well as personal questions. Will the conventional attempts to conserve habitat through parks and wilderness preserves and zoos actually stop extinctions, or are these strategies really just stopgaps, leading to such false triumphs as slowing down the rate of increase in the number of extinctions? Can high-tech biology pull us through? And will we grow content with our diminished lives, losing ourselves in football games on TV or golf on Saturdays as the apocalypse draws closer?

Yet something else has happened with the rise of endangered species in North America: We now have a totally new, totally modern category of animals—shadowland creatures, neither certainly extinct nor certainly living. Not all species are as obliging to us as the passenger pigeon, the Carolina parakeet, or the dusky seaside sparrow—the last ones dying before our eyes in zoos where their loss can be certified and documented. Many species vanish quietly. Most of the species that have vanished since 1940, such as the ivory-billed woodpecker of the southeast, have not been declared "officially" extinct. They have no real status as either extinct, to be mourned, or endangered, to be protected. This leaves biologists—and the rest of us who want our world carefully defined and categorized—with troubling paradoxes: How do you document what is not there? How do you prove an extinction? How do you live in the shadow of what neither is nor is not?

Several creatures inhabit this shadowland, and they offer the perfect metaphor for what life has become for us. The Eskimo curlew, a large sandpiper with a long, downward-curving bill, is extremely rare, almost extinct: The few records from this century come from along the Texas coast, and the bird has inspired searches recently for nesters on the arctic tundra of Alaska. The Bachman's warbler, the rarest warbler in the United States, is a small yellow bird from the southeastern swamps—it may already be extinct. The West Indian monk seal was discovered by Columbus in 1493; hunted relentlessly, the last recorded West Indian monk seal was killed in U.S. waters off Key West in 1922. The species is probably extinct, though quixotic hopes persist. The North Pacific right whale has a similar status, and a similar allure. The most famous animal in this ambiguous noncategory is the ivory-billed woodpecker. It was the biggest of North America's woodpeckers, and the second largest in the world. (Only Mexico's imperial woodpecker was bigger, and it, too, may now be extinct.)

I have always been drawn to the quest for the ivory-billed woodpecker, called in Latin *Campephilus principalis*—something about the combination of the bird's rarity and the hopelessness of actually finding one. There has not been a confirmed sighting of the ivory-billed woodpecker in almost half a century, despite intense efforts and fierce battles over purported sightings, and the rarer the bird has become, the greater its appeal. Woodpeckers have always been one of my favorite birds—I love their energy, their ebullient and undulating flight patterns, their raucous cackling that enlivens a forest, their habit of beating their heads against hard trees. And the ivory-billed is the preeminent endangered woodpecker.

The ivory-billed woodpecker was the most beautiful of North America's woodpeckers. A magnificent bird, it was easily able to occupy a quasi-mythical status: a bold clean statement in black and white and burning red. As big as a crow, it had a disproportionately large bill, off-white and extremely powerful. The ivorylike bill evokes associations with valuable tusks of other rare and appealing creatures such as the elephant, the walrus, and the narwhal. Its dramatically patterned plumage gave it an assured air of aristocratic simplicity and elegance. Through its face, a single white stripe flowed back from its bill and down its neck, like a satin scarf over a black tuxedo. The white stripe poured into a pool of white feathers on its folded wings, a big, distinctive triangle in brilliant contrast to its glossy black body. On the female, the large crest was pointed and black; when erect on the bird's head, it suggested woodpecker energy and woodpecker intensity. On the male, the crest peaked above the head in wild scarlet, like the flame of a fiery anointment.

Most searches for the ivory-billed woodpecker in the United States now begin in the Atchafalaya Basin of Louisiana, just west of Baton Rouge. Several of the most promising recent sightings come from the Atchafalaya swamp. It is 17 miles wide and 60 miles long—some 700,000 acres of still water and barred owls, armadillos and cypress trees, mud and scum. The largest river swamp in North America, it also offers the largest hope that ivory-billed woodpeckers may actually survive in the relatively unfrequented recesses.

In 1986, I spent almost two weeks in the Atchafalaya swamp, camping on a piece of high ground, owned by Gene Davison, about 2 miles upstream from the Upper Pigeon Landing on the Upper Pigeon River, just west of Bayou Sorrel. The location was recommended by Bruce Crider, an expert amateur ornithologist who claimed to have seen and heard an ivory-billed woodpecker there in 1978. Gene Davison, a retired firefighter from Baton Rouge, is by his own description a "good ol' boy," big chested and full of pranks. He liked to tell stories. He retired with his wife, Hazel, to a faded house along the river, not much more than a shack really, where he catches crawfish and catfish and jokes with the Cajuns about "how po' they is." One of the guys who gathered in afternoons on Gene's porch for community coffee told me he was "so po' his mammy had ta stuff the turkey with newspapers." Most of the guys I met on Gene's porch claimed to have seen lots of ivory-billed woodpeckers in the Atchafalaya.

Gene was my guide in the labyrinths of the swamp. One day, he took Bruce Crider and me in a boat through the Atchafalaya Flats, a vast laky swamp. Since his disputed sighting of the ivory-billed on Gene's property, Bruce has kept active in the search for the bird, becoming a sort of amateur expert on the politics and biology of ivory-billed woodpeckers. Gene wanted to show us a big woodpecker hole in a cypress, one he thought might be from an ivory-billed woodpecker, which he called a "log god," one of the common names for the big woodpeckers of the south. We wandered among huge cypress trees rising out of the water, scalloped at the base where the roots form just above the surface of the water, looking like knuckles. The reflections of the cayenne-colored trees shone off the black water in upside-down, disorienting images. Gene took us through a broad, open section of water and then into a confusing maze of denser cypresses. Blue and white and orange plastic tags marked the locations of crawfish traps, tracing routes through the swamp, but I was lost. I never would have found my way out of there alone.

When we got to the tree we were looking for, Bruce claimed the hole was too small and round for an ivory-billed woodpecker. It was probably a near cousin, the much more common pileated woodpecker.

Drifting through the swamp, listening to Bruce tell ivory-billed woodpecker stories, I let myself get dazzled by the reflections of vines and tree trunks in the water, not really expecting to see ivory-billed woodpeckers. As Bruce talked of his sighting and his conviction that ivory-billed woodpeckers still live in the Atchafalaya, I began to feel that words were all we have left, faint trails of speech that are their own empty reflections of the birds. The bird lives largely in our language, in our stories, which are now a thin echo of its absence.

It's a word trail that goes back at least fifty years, to when James Tanner conducted the only study ever made of ivory-billed woodpeckers. He published his conclusions in 1942 in a monograph, *The Ivory-billed Woodpecker*, which, if not definitive, is the authoritative account of the bird's biology. As early as 1920, the bird was feared extinct, largely through the cutting of its bottomland hardwood forests. When Tanner studied the species, he concluded that there were between five and twenty-two ivory-billed woodpeckers in Louisiana. After searching the south, he could find no evidence of the birds anywhere else. Since then, we have had only rare reports of the bird. The following list, compiled by the U.S. Fish and Wildlife

Service, cites the more-or-less creditable reportings. The U.S. sightings since the early forties have not been confirmed, many have been hotly disputed or discredited, and the list stands as an expression of the bird's fading elusiveness.

1933, December 25—north Louisiana: George Lowery, the patriarch of ornithology in Louisiana, saw a pair of ivory-billed woodpeckers in the Tensas area.

1935, April—Tensas River area, north Louisiana: George Sutton made pencil sketches of a live pair of ivory-billed woodpeckers.

1935—Santee River Swamp, South Carolina: Alexander Sprunt, Jr., guided Lester L. Walsh of the National Audubon Society through the swamp. They saw one ivory-billed woodpecker and heard two others.

Late 1930s—Singer Tract, north central Louisiana: James Tanner, Arthur Allen, Paul Kellogg, and George Lowery took photos of ivory-billed woodpeckers and made the only verified recording of its call. By 1948, the Singer Tract had been extensively logged over.

1941, December—Singer Tract, north central Louisiana: James Tanner last saw ivory-billed woodpeckers here. They were two females, one of which was a young bird raised that year.

1942, May—Singer Tract, north central Louisiana: Roger Tory Peterson caught glimpses of two birds. This may be the last authoritative, or officially recognized, sighting.

1950, March 3—Chipola River Swamp, northwestern Florida: Whitney H. Eastman saw a male ivory-billed woodpecker.

1950, March 4—Chipola River Swamp, northwestern Florida: Eastman saw a female ivory-billed woodpecker.

1950—Chipola River Swamp, northwestern Florida: James Tanner and Bob Allen visited the swamp but could find no sign of ivory-billed woodpeckers.

1951, April 5—Chipola River Swamp, northwestern Florida: John V. Dennis heard an ivory-billed woodpecker.

1958—Altamaha River, Georgia: From a small plane, Herbert

L. Stoddard, a well-known ornithologist, saw an ivory-billed woodpecker in flight.

1958—near Thomasville, Georgia: Stoddard sighted a pair of ivory-billed woodpeckers.

1966, August 28—Boiling Creek, a tributary of the Yellow River, northwestern Florida: Bedford P. Brown, Jr., and Jeffrey R. Sanders, Chicago birdwatchers, heard the call notes and saw two female ivory-billed woodpeckers. They reported the sighting to John V. Dennis, who tried unsuccessfully to find the birds again.

1966, December 3—Neches River Valley, Texas (Big Thicket): John V. Dennis heard an ivory-billed woodpecker.

1966, December 10—Neches River Valley, Texas (Big Thicket): John V. Dennis saw an ivory-billed woodpecker.

1967, February 19—Neches River Valley, Texas (Big Thicket): John V. Dennis and Armand Yramategu saw an ivory-billed woodpecker.

1967, April to early June—Neches River Valley, Texas (Big Thicket): John V. Dennis was hired by the Bureau of Sports Fisheries and Wildlife to look for the ivory-billed woodpecker. At the end of May, he estimated five to ten pairs in the Neches Valley.

1968, January—Neches River Valley, Texas (Big Thicket): Paul Sykes, of the U.S. Fish and Wildlife Service, and James Tanner searched the area to confirm Dennis's sightings, but they could find no evidence to indicate that the species was or had been in the area in the previous several years.

1968, February 25—north of Beaumont, Texas: Dennis recorded what he believed to be an ivory-billed woodpecker. The tape was analyzed by Peter Paul Kellogg at Cornell University and by John W. Hardy at the Florida State Museum, University of Florida. Both thought it sounded like the call of an ivory-billed woodpecker.

1968, April 21—Polk County, south central Florida: H. Norton Agey and George M. Heinzmann collected three feathers and made a tape of a calling bird. They also published several sightings of the ivory-billed woodpecker between 1967 and

1969. Paul Sykes went into the field with these men and, in one report, "was not impressed." James Tanner listened to the tape, and he said nothing on it could be an ivory-billed woodpecker.

1968 or 1969, Fall—Navasota River, Texas: Robert Lys, a game management officer, claimed to have seen an ivory-billed woodpecker.

1970, October 30 and December 24; 1971, January 8—Sam Houston National Forest, Texas: Bill Ruediger, a technician with the Forest Service, reported seeing ivory-billed woodpeckers. The sightings were not confirmed.

1971, May 22—Atchafalaya Swamp, Louisiana: An unidentified dog trainer saw a pair of ivory-billed woodpeckers and photographed a male. He sent the photographs to George Lowery, who made several trips to the area but could not find the birds. Bruce Crider found out who took the photographs, and he talked to the man. These pictures have been hotly disputed.

1973, March—near the confluence of the Tombigbee and Noxubee rivers, Alabama: Jerome Jackson, woodpecker biologist, observed a large woodpecker with extremely large white wing patches, like an ivory-billed woodpecker. It flew across the river while he was canoeing.

1973, December 8 and 9—Neches River Valley, Texas (Big Thicket): U. L. Loucks saw considerable bark stripping, a sign of ivory-billed woodpeckers. Tanner and Sykes investigated, and they said the stripping was caused by pileated woodpeckers.

1974, November 11—Atchafalaya Basin, Louisiana: Robert Bean, director of the Louisville Zoo, reported seeing an ivory-billed woodpecker fly across Interstate 10, less than 15 feet away from him.

1976, May 22—Big Thicket area, Texas: William Mounsey of the University of the Wilderness in Evergreen, Colorado, reported an ivory-billed woodpecker.

1976 or 1977—Atchafalaya Basin, Louisiana: Erica Tallman saw an ivory-billed woodpecker below her on the Interstate 10 bridge across the swamp.

1978—Atchafalaya Basin, near Upper Pigeon Landing, Louisi-

ana: Bruce Crider saw and heard an ivory-billed woodpecker while he was playing a tape.

1981, April 10—Atchafalaya Basin, near Bayou Chene, Louisiana: Thomas Michot and David Hankla, biologists with the U.S. Fish and Wildlife Service, reported seeing an ivory-billed woodpecker, which landed in a willow tree ahead of them. They heard the bird call. Michot also photographed the bird, though a large leaf blocked most of the bird's body in the photograph. The Ivory-billed Woodpecker Advisory Committee of the U.S. Fish and Wildlife Service examined the photo and said "it is clearly of a pileated woodpecker."

1984, June 23—between Crosett and Hamburg, Arkansas: Dorothy Jenkins said she had seen an ivory-billed woodpecker fly across Route 7.

1987, March—junction of the Yazoo and Mississippi rivers, Mississippi: Jerome Jackson, of Mississippi State University, and Malcolm Hodges, one of his graduate students, had a bird respond repeatedly for twenty-eight minutes to a tape recording. They never saw the bird, but its call was indistinguishable from that on the tape.

1986 to 1988, Cuba's Oriente Province. Several researchers, most notably Lester Short of the American Museum of Natural History and Jerome Jackson of Mississippi State University, saw ivory-billed woodpeckers here, confirming that at least the Cuban subspecies survives.

This is it: the modern legacy of ivory-billed woodpeckers. In America, it is a long official series of unconfirmed reports. Tokens of hope on the one hand; empty words on the other. A shadowland between extinction and existence.

A French philosopher (Jacques Lacan, I think) has written somewhere that the birth of the symbol is the death of the fact. In the case of the ivory-billed woodpecker, the process has been reversed. As the bird became rarer, its already large reputation and allure grew: It has been transformed into a creature of mythic dimensions. As the ivory-billed slowly receded into the shadowy swamps of the south, both the bird and its status grew more elusive and indefinite. You can't see the bird, but if you watched carefully as you read the list above, you will have seen quite clearly its metamorphosis into metaphor, a stunning bird flying from fact into fiction. The history of the bird offers just this disquieting sense of dislocation and displacement. In a surreal way, reality and fantasy have grown radically dissociated.

What we have left is not the bird, not living flesh, not the flashes of power and brilliance that early naturalists reported, but a half century of purported and disputed sightings. Taken together, they create the eerie minimalist sense that all we have left are rumors of existence.

Creatures like the ivory-billed woodpecker have given rise to an utterly new, utterly modern phenomenon—a new kind of quest. In the Middle Ages, knights went on searches for fabulous creatures such as the unicorn or for fabled objects such as the Holy Grail. In these medieval quests, a whole culture signified its faith in the meaning of life—there was a power and a sacramental vision that could be achieved.

But the ivory-billed woodpecker and other shadowland creatures show the transformation of this quest. Look to what sad ends we have come. Like the unicorn or the Holy Grail, the ivory-billed woodpecker and the West Indian monk seal and the Eskimo curlew have exotic, romantic names—perfect for the quest and the test of our faith in life. Only now what? The quest is not for the meaning of life but for the almost or certainly extinct creature. Though these new creatures of the quest inspire profound faith in many people, who devote themselves and major parts of their lives to finding them, who become champions-errant of a nearly lost cause, the very meaning of these quests is defined by their near hopelessness. The quest takes its being from rumors, and it wanders in the region between what is and what is not.

A quest for the ivory-billed woodpecker is the perfect modern narrative: It takes place in a swamp, nothing happens, you're going nowhere, and there's probably nothing to find.

Listening to Bruce and Gene tell stories as we floated more or less aimlessly through the flats, I had a sense I sometimes get: Like the ivory-billed woodpecker, I sometimes seem to know that I exist only because I hear other people talking about me.

It was a beautiful April day when Gene led Bruce Crider and me into the Atchafalaya flats, and the sunlight lanced through the branches of cypress and tupelo trees. By the shore, sweet gum and laurel oaks showed the early green of budding leaves. Vividly yellow prothonotary warblers sang in the branches, and hooded warblers flitted away from us in the small hawthorne trees. We passed two barred owls, sitting like bookends on opposite sides of a bare tree. Cormorants spread their black wings in the sun to dry.

We did not find any signs that day of ivory-billed woodpeckers, other than in our talk.

II

Van Remson led us past several aisles of army-green cabinets in the darkened room, turned right at the far wall, and stopped purposefully at a drawer beneath a poster of Mexican hawks. Van is the curator of the Museum of Natural History at Louisiana State University in Baton Rouge. I met him before heading into the Atchafalaya, since he presides over much of the lore about the ivory-billed woodpecker in the state of Louisiana. Van is convinced there are ivory-billed woodpeckers in the Atchafalaya, and he showed me the two disputed photographs of the woodpeckers taken in 1970. (The birds looked like ivory-billed woodpeckers to me.) Short and slight, he walked with a limp, from a sprain in a racquetball game, which belied his reputation among ornithological circles for tough work in rugged tropical areas of Latin America. His long hair hung in wispy, blond strands.

Van pulled out the drawer and placed the skin of an ivory-billed woodpecker in my hand. On its leg tag, a handwritten script read "Tarpon Springs, Florida, 1887"—the specimen was collected over a century ago. All those years, lying in a museum drawer, had left it flattened on one side. It was late at night when I met Van, and the museum had a lonely, empty, hollow feel to it.

Then Van pulled the skin of a pileated woodpecker out of another drawer. We laid the two skins side by side. I had always thought that the pileated woodpecker was an unsurpassed woodpecker for its beauty. But next to the ivory-billed skin, it suddenly seemed a pretender. I will never see a pileated in the same way again. Like the ivory-billed woodpecker, the pileated was black and white and red. But its face, a male face, was more complicated in its markings—red crest, white stripe, black, white, red and black stripe, white chin. Next to the ivory-billed skin, its markings suddenly seemed too busy.

Though I had studied pictures of ivory-billed woodpeckers, nothing prepared me for this corpse. It had the force of a bold gestalt. The red crest of the ivory-billed woodpecker rose to a higher, finer point than the pileated's. It had a cocky, upswept aplomb, with a superb black border outlining the front of the crest with the hint of a forward flipping curl. The face didn't have the clutter of the pileated woodpecker's face, just a single white line through a black face, like a beam of light through darkness. The bill was much bigger and, even though yellowed with age, strikingly beautiful. And on its wings, not the hints of white shown on the pileated but bold, strong, emphatic triangles. Apparently, in flight, the large white swatches on the trailing edges of the ivory-billed woodpecker's black wings had a stunning effect.

A corpse as an epiphany. It seemed to me that night as if the idea of woodpecker had been taken to its essential form, a perfectly realized creation.

There are about 400 ivory-billed woodpecker skins in museums around the country. As the bird grew rare at the beginning of this century, scientific collecting actually contributed to its extinction.

But to get a feel for what this bird must once have been, to feel something of its power over the human spirit as a living creature, we can turn to earlier American naturalists. Mark Catesby, an Englishman who lived from 1683 to 1749, has been called "the founder of American ornithology," though his primary interest was botany. His reputation is founded on *The Natural History of Carolina, Florida, and the Bahaman Islands*, published in London in 1731, in which Catesby gave us many of the names that we still use today for

American birds, such as "The Blew Jay" and "The Blew Bird." His lasting contribution came, however, not as a cataloger of plants and animals but as a painter and descriptive naturalist.

In his portrait of "The Largest White-bill Woodpecker," Catesby offers a quaint bird, stiff and motionless, posed in rigid profile like a wide-eyed mannekin. The face and eyes are full of charm and childlike innocence. All of Catesby's portraits of birds betray an unassuming man, himself wide-eyed in the delight of first vision.

He says nothing in his text explicitly about his experiences with the ivory-billed woodpecker, but his much-quoted passage of the bird's importance to American Indians offers a clue to its perceived value:

> The bills of these birds are much valued by the Canada Indians, who make coronets of them for their princes and great warriors, by fixing them around a wreath, with their points outwards. The northern Indians having none of these birds in their cold country, purchase them of the southern people at the price of two, and sometimes three buck-skins a bill.

This is practical prose, the simple rendering of information. Yet it contains the poetry of clear perception. The ivory-billed woodpecker was worthy of princes and warriors, its regality more than metaphorical. An ivory-billed woodpecker's beak has been found in an Indian grave in Colorado. Also, several Indian pipes in the Milwaukee Public Museum are decorated with bills and feathers from the ivory-billed woodpecker; these were used in ceremonies by Iowa Indians.

In the next century, inaugurating the golden age of American ornithology, Alexander Wilson (1766–1813) published an unprecedented collection of paintings and stories of American birds, *American Ornithology*. The first volume appeared in 1808. A weaver, a schoolmaster, a poet, a lover of American ideals, a self-taught naturalist, Wilson set a new direction for the understanding of American nature. He is commemorated in the names of many of our birds, including the Wilson's warbler. His work shows the changing times, its exuberance and romanticism contrasting with Catesby's reserved realism. Wilson's description of the ivory-billed woodpecker exudes a love of liberty and a sense of the power of nature:

> . . . the royal hunter [of trees] now before us, scorns the humility of [stumps and posts], and seeks the most towering trees of the forest; seeming particularly attached to those prodigious cypress swamps, whose crowded giant sons stretch their bare and blasted or moss-hung arms midway to the skies. In these almost inaccessible recesses, amid the ruinous piles of impending timber, his trumpet-like note and loud strokes resound through the solitary, savage wilds, of which he seems the sole lord and inhabitant.

Wilson's inflated prose turns the ivory-billed woodpecker into a romantic hero, a Childe Harold of the American swamp. The ivory-billed woodpecker also embodied Wilson's fervent democratic ideals.

He also tells a personal anecdote of the ivory-billed woodpecker, more intimate and affecting, trying to capture what he calls "the disposition of the bird." It is the story of his first encounter with an ivory-billed woodpecker—the bird he painted and included in his plates of American birds. Wilson found it about 12 miles north of Wilmington, North Carolina. The bird was wounded slightly in the wing, and Wilson caught it, whereupon it

> uttered a loudly reiterated and most piteous note, exactly resembling the violent crying of a young child; which terrified my horse so, as nearly to have cost me my life. It was distressing to hear it.

Wilson carried the bird with him to Wilmington, where its "affecting cries surprised every one within hearing, particularly the females, who hurried to the doors and windows with looks of alarm and anxiety."

Wilson rode to his hotel and asked for an accommodation for himself and his "baby." The attendants stared at him in astonishment.

Taking his woodpecker with him into his room, Wilson left the bird alone for a few hours. He returned to find a screaming woodpecker:

He had mounted along the side of the window, nearly as high as the ceiling, a little below which he had begun to break through. The bed was covered with large pieces of plaster; the lath was exposed for at least fifteen inches square, and a hole, large enough to admit the fist, opened to the weather-board; so that, in less than another hour he would certainly have succeeded in making his way through. I now tied a string round his leg, and fastening it to the table, again left him. I wished to preserve his life, and had gone off in search of suitable food for him. As I reascended the stairs, I heard him again hard at work, and on entering had the mortification to perceive that he had almost entirely ruined the mahogany table to which he was fastened, and on which he had wreaked his whole vengeance. While engaged in taking the drawing, he cut me severely in several places, and, on the whole, displayed such a noble and unconquerable spirit, that I was frequently tempted to restore him to his native woods. He lived with me nearly three days, but refused all sustenance, and I witnessed his death with regret.

Under the romantic eulogies for the bird's love of freedom and the wild, what Wilson sees in the ivory-billed is irrepressible personality, even as he watched it die.

He reports also that southern Indians held the ivory-billed woodpecker in "great esteem," using the head and bill as amulet and ornament: "And as the disposition and courage of the ivory-billed woodpecker are well known to the savages, no wonder they should attach great value to it, having both beauty, and, in their estimation, distinguished merit to recommend it."

Thomas Nuttall (1786–1859) issued *A Manual of the Ornithology of the United States and Canada* in 1832. Writing in the tradition of Wilson, reflecting romantic and democratic values, Nuttall says that the Indians of the south admired the ivory-billed woodpecker because it showed "the unconquerable spirit of a genuine son of the forest":

From his magnanimous courage and ardent love of liberty, the head and bill are in high esteem among the amulets of the southern Indians.

Nuttall's prose also conveys the effect that "this princely wood-pecker" has had on people:

> His retiring habits, loud notes, and singular occupation [pounding on trees] amidst scenes so savage yet majestic, afford withal a peculiar scene of solemn grandeur, on which the mind dwells for a moment with sublime contemplation.

John James Audubon (1785–1851) transforms the romantic exuberance associated with the ivory-billed woodpecker into a fresh and vivid theatricality. Writing as a man who tried to embody the American frontiersman, he brings his own posturing personality to his confrontations with nature. With his sweeping, emotional gestures and his love of the grand statement, he found in the ivory-billed woodpecker an excellent vehicle for the exercise of his imagination. In Audubon's descriptions, art meets nature. In his *Ornithological Biography*, published in five volumes between 1831 and 1839, he transforms the bird into the embodiment of cavalier beauty and the natural sublime:

> I have always imagined, that in the plumage of the beautiful Ivory-billed Woodpecker, there is something very closely allied to the style of colouring of the great VANDYKE [sic]. The broad extent of its dark glossy body and tail, the large and well-defined white markings of its wings, neck, and bill, relieved by the rich carmine of the pendent crest of the male, and the brilliant yellow of its eye, have never failed to remind me of some of the boldest and noblest productions of that inimitable artist's pencil. So strongly indeed have these thoughts become ingrafted in my mind, as I gradually obtained a more intimate acquaintance with the Ivory-billed Woodpecker, that whenever I have observed one of these birds flying from one tree to another, I have mentally exclaimed, "There goes a Vandyke."

The locations where he finds the woodpecker are part of the appeal, part of the act:

> I wish, kind reader, it were in my power to present to your mind's eye the favourite resort of the Ivory-billed Woodpecker. Would that

I could describe the extent of those deep morasses, overshadowed by millions of gigantic dark cypresses, spreading their sturdy moss-covered branches, as if to admonish intruding man to pause and reflect on the many difficulties which he must encounter, should he persist in venturing into their almost inaccessible recesses, extending for miles before him, where he should be interrupted by huge projecting branches, here and there the massy trunk of a fallen and decaying tree, and thousands of creeping and twining plants of numberless species. Would that I could represent to you the dangerous nature of the ground, its oozing, spongy, and miry disposition . . . that endangers the very life of the adventurer, whilst here and there, as he approaches an opening, that proves merely the lake of black muddy water, his ear is assailed by the dismal croaking of innumerable frogs, the hissing of serpents, or the bellowing of alligators!

Audubon has made nature into a stage for his own self-dramatization. For him, nature is nothing less than an imaginative experience in which he creates himself in the romantic American image.

The result is that he projects exaggerated vitality into his paintings. The three ivory-billed woodpeckers in his painting strike electric poses as they focus with concentrated energy on a single, slightly off-center point of convergence—a beetle. The vigor of the painting, especially in the two birds in front and back poses, conveys itself in the pointed beaks, the bent and twisted bodies, the wings and tails flexed in muscular tautness, the single flying piece of bark.

The lore of the ivory-billed woodpecker is immensely compelling. A bird revered for its majestic beauty and fierce spirit, it was a creature with personality, closely associated with a young nation's sense of itself. Audubon's swamps of the south, for example, suggest a nation with untold numbers of trees and plants, a nation of inaccessible recesses we can never fully explore, a nation of alligators and hissing snakes and innumerable frogs. The images of the ivory-billed woodpecker which we have inherited from the early naturalists are seductive. I can easily wrap myself up in them and give myself over to the fantasies of the past.

Except I also remember the corpse in my hand. The ivory-billed woodpecker took on a tangible existence for me in that skin. It had weight and substance. I could hold it. It was exciting to see, evoking a sense of what the bird once was. But it was also an intrusive rebuttal: a beautiful corpse, with white cotton poking out of empty eye sockets.

The ivory-billed woodpecker as a creature is not the only thing we have lost. We have also lost the possibility of the fantasy Wilson and Audubon created out of the American swamps. The unconquerable spirit that Wilson saw in the ivory-billed woodpecker was quickly vanquished—the bird was virtually gone before the nineteenth century was out. Audubon's vast swamps live on in relics: The Atchafalaya looks wild, for example, but it is completely controlled by the Army Corps of Engineers—its flow regulated, its perimeter surrounded by levees. And with each new extinction, with each new diking of a swamp, with each newly cut forest, some part of the country inside dies too. The fantasies of the past are irrelevant, anachronisms in the world that has destroyed them. The corpse in my hand is what I have to live with.

III

Camping in the Atchafalaya, looking for ivory-billed woodpeckers I didn't expect to find, I spent days hiking through mud and seeing everything but what I most wanted to see. In the process, I spent a lot of time in my head, and I got lost in my several selves. One self I ran into there was the one that makes me most uncomfortable, the one I know least how to show publicly—he wears rags, isn't sure where he's going, hardly knows where he's been. He's one of my shadows. He scares me, because when he appears, I start to vanish. Like the ivory-billed woodpecker, he lives close to extinction, amid rumors of survival.

My camp was on a bank above the Upper Pigeon River, where Gene Davison had a hunting camp. Trails led back into the swamp, well-worn from three-wheelers Gene and Hazel used to get to the standing water where they set their crawfish lines. One morning I walked straight back into the swamp on one of the trails, playing a copy of the ivory-billed woodpecker tape made in the 1930s by Dr. Arthur Allen in northern Louisiana. I had little confidence in finding an ivory-billed woodpecker; I had been at this spot for several days with nothing to bolster even the slightest hope. I also had this weird guy inside me that I had to keep busy, entertained, distracted. So I hiked.

The swamp was Edenic in a hellish sort of way. Where the ground was unflooded, fields of waist-high flowers called yellowtops lit up the forest floor as if it were a lower sky, and I loved walking through them off the trail, listening to the slurred songs of warblers floating down from the oaks. Budded leaves on the trees curled in flecks of April green. Almost imperceptibly, the ground sloped down toward the swamp, and about a mile back from the river, the yellowtops gave way to exposed mud. The trail grew faint and in several places ran into orange pools of water, big as a city block. Skirting these waters, I squished through mud up to my ankles, looking for the highest ground possible, listening for woodpecker calls.

Woodpeckers were everywhere. Their calls clattered from every direction, like pots banging in a kitchen. Flickers: common throughout the country. Yellow flashes under their wings as they fly. They swooped suddenly onto a tree, not so much landing as sticking to the trunk with the abruptness of a flung knife. They blurted out a raspy call—a single, loud scream like the cry of a frightened crow. Downy woodpeckers: the smallest of the woodpeckers, only 7 inches long. They have small beaks for woodpeckers. I loved to watch them hopping up the side of a tree, using their stiff tails for support, always busy, their lilting call like a soft whinny. Red-bellied woodpeckers: abundant in the Atchafalaya and throughout the swamps of the southeast. Black and white streaks give the red-belly a ladder-back, and the male has a vermilion cowl from beak to back. It, too, had a chattering, bouncing call.

Reports vary on how abundant ivory-billed woodpeckers once were. Audubon suggests they were everywhere in the southern swamps, but they may never have been very common. For one thing, they were specialized creatures. Tanner concluded that ivory-billed woodpeckers required virgin forests, where they could find large numbers of newly dead trees. Their huge bills were an adaptation to this habitat: Most woodpeckers pound into a tree, but the ivory-billed woodpecker used its bill as a chisel rather than a drill. The ivory-billed would knock big chunks of bark off newly dead trees with its blunt beak, looking for borer larvae. Big tracts of old trees were needed to supply enough of these larvae. As the old trees in the swamp bottoms were harvested, the ivory-billed woodpecker vanished.

While I played my tape, I looked for trees with patches of stripped bark. Of course, I found none, but I kept looking anyway.

The ivory-billed is probably extinct in the United States. There remains a lot of debate about the issue, but even that debate is dying out. In fact, the debates over whether the bird still existed conceal our failure to help this woodpecker. Though it is without question one of the most endangered and well-known animals in North America, we have spent almost no money to save it. Since we couldn't prove it existed, it wasn't an official problem. As Marshall Jones, the chief of the Endangered Species Division, Southeast Regional Office, in Atlanta told me, "Is the ivory-bill a problem? No. Not really. A problem demands you do something about it. We haven't done anything. If we find 'em, then there's a problem."

In 1987, Jones's office funded a two-year study by Jerome Jackson to search the swamps in the south for ivory-billed woodpeckers. A preeminent woodpecker biologist from Mississippi State University, Jackson began his quest with an amazing confidence, sure that he could document the existence of ivory-billed woodpeckers. By the end, his hope had faded. In his two years in the swamps, his only sign was one call that he thought was an ivory-billed woodpecker. Yet he remains committed to the bird: "I still believe," he told me.

There was another brief flurry of hope for the ivory-billed woodpecker in 1986, when Lester Short of the American Museum of Natural History traveled to Oriente Province in remote eastern Cuba. Cubans had reported sightings of ivory-billed woodpeckers, and Lester Short found the birds and broke the story to the press. Cuba is the only other place, besides the southeastern United States, where the ivory-billed woodpecker occurred.

In 1988, Jerome Jackson made the trip to Cuba. Standing on a cliff on March 4, he got a three-second glimpse of an ivory-billed woodpecker as it flew 30 feet below him—the only certain sighting he's had in his life. Though the place was remote, Jackson said the area was already heavily logged, and both mining and agriculture are moving in—bad signs for the species' future.

"We overestimated what might be left in Cuba," Jackson told me. "Maybe a handful of ivory-bills are left there. They could be down to their last two or three. There's more hope in Cuba than there is here now, but only because they've been seen there. The species may be really on the way out. By the end of the century, it will be gone."

I edged my way up a small hummock covered with grass and lichen-mottled bushes, along a ditch of stagnant black water. Clumps of green algae and a film of pollen floated on the surface. While I was watching a waterthrush flitting near the shoreline, I noticed a cottonmouth snake beneath the tangled branches of a bush. Unmoving, it lay in the water like just another stick, smoky gray, with a diamond head—the sign of a poisonous snake. The waterthrush flitted dangerously low, just above the cottonmouth, and I thought I might see the snake strike.

The cottonmouth reminded me of all the other creepy animals I'd seen in the swamp. One night, fishing for catfish from a muddy bank, I caught a slithering green eel that I could hardly get off the hook. Checking a fish line with Gene on another night, we'd pulled up a strange paddle-nosed fish. Once, as I was stomping through the swamp with Gene, helping him work his crawfish route, we'd found a big blue salamander, a foot long, with ridiculously small legs and small feathered gills. Another time, sitting at my campfire during the night, I heard an immense crashing in the bushes, as if a big bear were coming through, only to watch a little armadillo pop out into the circle of light around the fire.

While I was absorbed with the cottonmouth, chilled by the Jurassic feel of the swamp, a loud and crazy laughter upset the stillness of the morning. From the opening along the backwater, I looked up to see two large woodpeckers fly over the canopy and careen into the bare branches of a tall tree not yet in leaf. With each beat of their wings, white blazed against their black bodies. White that was only on the leading edges of their wings. No white on the trailing edges, which was a field mark of ivory-billed woodpeckers. These were pileated woodpeckers, cackling with wild, nearly hysterical calls.

It was disappointing. I tried in my mind, for a moment, to pretend they were the rare ivory-billed woodpecker.

Pileated woodpeckers are common birds, almost abundant if you know where to look for them, and what I regretted was their commonness. They were once much rarer, reaching a low point around the turn of the century. But they apparently learned to adapt to less primitive conditions as the forests were cut, and around 1920 they began a comeback. They can now be found in parks and woodlands as well as deep woods. What makes me sad about the pileated woodpecker is the phenomenon it represents in our national wildlife: Like the pileated and the ivory-billed woodpeckers, many animals seem to come in twos, near relatives to each other. Under the pressures of human expansion, the more spectacular of the two species dies out or grows rare, like the ivory-billed woodpecker, while the other either adapts or is tolerated, like the pileated.

Think of whooping cranes and sandhill cranes. The tallest bird in North America, whooping cranes are dramatic white creatures, 4½ feet tall, with red patches of skin on their heads. Never abundant, their population fell to a dismal low of twenty-one in 1944 as they lost their marshes to agricultural development. Intensive management has brought that number up to about 134 wild cranes now, most of which winter in Aransas National Wildlife Refuge on the Texas Gulf Coast and nest in northern Canada. Another small flock has been created in Gray's Lake, Idaho, where biologists used sandhill cranes as foster parents for whooping crane eggs.

The sandhill crane is also large, a gray bird that is much more common and adaptable, with several hundreds of thousands.

Think too of trumpeter swans and tundra swans. The trumpeter swan is the biggest waterfowl in the world, large and white and elegant. Once widely distributed over North America, it fell to as few as seventy birds—the only known remnants of the species living in the remote mountains of extreme southwestern Montana in the early 1930s. Because of intensive management, its numbers have been brought back to about 500 in the contiguous United States. As many as 10,000 have been discovered in the last three decades, but these birds are in Alaska, and they have returned in such spectacular fashion not because of management but probably because of the warming of the earth.

Tundra swans have remained much more common, with a population of about 100,000.

Or think of grizzly bears and black bears. The huge and hump-backed grizzly was once found throughout the western United States. But as humans pushed into its range, it was killed off, trapped and poisoned and shot because it was so fearless and fearsome. Currently, its range is limited to four northern states, and its population is estimated at 600 to 900 animals. It survives mostly in national parks, and there are still outcries in favor of killing it off.

The black bear is one of the most adaptable of the large carnivores in North America. Though it, too, frightens people, it has a population of better than 200,000 animals in the lower forty-eight states.

Think of spotted owls and barred owls. Needing old-growth forests, spotted owls are vanishing as the huge trees in the Pacific Northwest are cut, while the adaptable barred owl is extending its range into the west, taking over in the degraded forests that remain.

The adaptable and nonthreatening animals survive. Though they are not invulnerable to the pressures of humans, they adapt and manage to make do. But there is already too much commonness in our world, too much of the mundane. And there is something of classic tragedy in the other creatures, the rare and more spectacular ones, the ones we have destroyed. Rather than compromise, unable to adapt to new conditions imposed on them suddenly and violently, they die off. I find a kind of integrity in their inflexibility. They are what they are. And they are what we are killing off—variety, uncommon beauty, and integrity of being.

I don't want more commonness. I want more of the rare. These are the creatures that speak to the fragile part of myself, the part threatened by the demands of a world full of standardization and conformity. Commonness is its own kind of minimal existence. What scares me most is that we will—that I will—grow to accept commonness as one of the terms of existence. The greatest danger is that there will seem to be no alternative to our conventional lives. Or that we will have marginalized diversity to such an extent that it will survive only in small, well-defined pockets: in national parks for rare animals, or in rock stars for human beings.

The pileated woodpeckers flew out of their trees and farther into the swamp. I decided to go after them. They kept up their loud and raucous calls, and I followed their noise. The brown mud turned ocher, and I sloshed into water the color of rust, ankle-deep and stretching out of sight amid the cypress trees. The mucky water slowly deepened, until I was in over my knees.

They had landed on the trunk of a ancient bald cypress about half a mile out in the water. The tree's fibered bark of burnt orange blended with the tones of the swamp water, and its spreading and grooved base, where it entered the water in roots, reminded me of an elephant's foot. The two pileated woodpeckers were hammering on opposite sides of a dead branch high in the tree, the male just above the female. Chips flew from their beaks and rained into the swamp, and the birds worked with such energy that they didn't seem to be trying to accomplish anything. They seemed instead to be playing some game, harassing each other, using their beaks to rattle off threatening messages. They bent their heads this way and that, and they pounded into the branch from various angles, their dark beaks more like weapons than tools. Then the male started chasing the female around the branch. They hopped and shinnied, their talons clutching the wood, their crested heads and sharp beaks akimbo, their heads jerking frantically with each shinny. Their eyes glared with the intensity of the insane. They scolded each other, chased each other around the branch, and shinnied up higher, looking as gawky as a pair of teenagers climbing a rope.

At one point, the male scooted around the branch, screaming at the female, and the sun lit his crest from behind. In that brilliant red I could have sworn his head had caught fire and was burning in the trees.

Those two woodpeckers jumped and screamed and tattooed and

flew with a wild energy, as if possessed. Their crests erect, they jumped up the trees in a hysterical ecstasy: in the madness of a crazed woodpecker, pounding senselessly on a tree, not for food, but for who knows what? Announcing his presence to the swamp? For pure joy and freedom? A burning woodpecker screaming out his life in the trees.

There is an ancient story about woodpeckers, which the Roman Pliny the Elder tells in his *Natural History* (completed in A.D. 77). It is the most important piece of folklore about woodpeckers in the Western tradition. The woodpecker had magical power, freeing power. It had prophetic powers, for Pliny. If its nest hole got plugged, he reported, it would fly away and then return with the herb of life in its powerful beak; with the herb, the plug would magically fall out—a story of healing, and opening up, and trans-formation.

A breeze stirred the cypress leaves. If pileated woodpeckers are not as spectacular as ivory-billed woodpeckers, I can nevertheless love them, finding also the consolation that one being takes from another. And for myself? Whatever I am, in my woodpecker. soul, I have myself. I have what I see, and what I feel, even if it is not always what I have wished for.

What is there to do for the animals? The same thing I would do for myself: Love what's left, and fight for more. I was happy watching those pileated woodpeckers: knee-deep in swamp water, filled with a wild freedom, my mind on fire, burning like the wood-pecker's crest.

TEN

THE BACK DOOR

About five years ago, I realized that the windows in my house were broken. The wind poured in at night and rattled the pottery in the kitchen. The strange thing was that I had thought the house was secure. One evening, I rose from the dinner table and wandered out the back door. I have not found my way home yet. Instead, I have been walking through woods and back alleys, looking into the night and shadows, chasing after the animals that have eluded me.

What I have seen, out the back door, are broken creatures, speaking of broken dreams. I have looked into the eyes of many of them and have seen, inevitably, reflections of myself. For myself and for them, the same terms of life apply. It is a hard notion for me to keep before my mind, though, because it seems to contradict all the trends. But as humans appear to grow stronger and to dominate the earth more completely, both in sheer numbers and in technological power, the animals grow weaker, diminished and displaced. What is true for them of physical existence is true for us in the spirit. The simple and fundamental fact of existence is increasingly the issue for the animals—physical existence. For me, the struggle is similar, but only occasionally is the question of physical being at issue. What is rare for me, what seems threatened and endangered, is to live out of a sense of my own being.

Broken creatures—they are not external problems. They cannot be handed over to agencies or specialists while we continue to insist on doing all that we have done to break these creatures in the first place. Broken creatures—they come down finally to each one of us. It comes down not even to what we do: cutting trees, or draining swamps, or shooting predators. It comes down to what we are, because how we treat the creatures depends on how we see them. The diminished and marginalized animals are a reflection of our diminished selves, lost in the clutter of our prosperous lives. Try as we might, we cannot separate the animals from us. Even their extinctions come back upon us, like a growing void. Perhaps at some point the earth will be so bare and depleted that we will no longer be able to ignore the consequences of our lives. Instead, now, these lost animals are the outward images calling us to an internal struggle for new hearts and new minds.

Tidy houses, two-career households, double-pane windows, well-lit rooms—meanwhile, in the basement, or out behind the garage, the corpses pile up. I hear about the lost animals on TV. I feel badly for them, I even try to help them. I join organizations like Greenpeace, and I send in my contributions to National Geographic. Those are important things to do, necessary even. But then I go back to sleep, not even realizing I have come to think somehow that the diminished beast is the real beast.

Still, like Hamlet, I could be content in my nutshell, except that I have bad dreams. Animals come to me in the night with strange messages—whales swimming in my blood, alligators luring me onto lonely peninsulas, where they leap on me from behind. I dream of tigers lurking in my basement, longing to be seen, longing to be listened to, longing to be free. One tiger showed me open doors and told me not to be afraid. I woke from my sleep excited and shaken.

Open doors, broken windows, strange speaking animals.

So I left through the back door. It's the only way out, if you want new vision. Or if you want to find the animals that are the living analogies of the heart.

What I have witnessed is a new wave of extinction. In many of the creatures I went to see, I realized recently that I have done what can no longer be done. The dusky seaside sparrow is now extinct. The last pure dusky seaside sparrow, Orange, died not long after I stood with him in his cage, an old sparrow flitting in the grass and sand with no place else to go. Even the genes of the dusky have vanished, since the crossbred Scott and dusky seaside sparrows were lost in a March 1989 windstorm. Our best efforts for them came to nothing. I picture the duskies clinging to the wire of their cages, looking around, though my memories of them feel like pitiful epitaphs for an extinct race. The last wild California condor has been captured and now lives in a zoo, and I live with my memories of flying alongside that bird and his mate up the Grapevine to their roost in the mountains, evening coming on. The fierce green eyes of the last black-footed ferrets continue to haunt me, and like the condors, the ferrets now live in cages, their babies born in boxes, and you can no longer see them in the wild. Florida panthers are so endangered that they, too, may soon be captured, if they don't go extinct first. The panther I saw in the swamp was captured in 1987, in terrible shape, and died in August 1988.

And this whole list is summarized in the ivory-billed woodpecker, which may have gone extinct long ago, or may be going extinct now. We simply don't know, which is a kind of appropriate mockery of our pride and of the arrogance that informs so much of our knowledge. Has it gone extinct? Somehow, being able to define the loss, to give an exact date to our failure, would be no consolation.

These creatures are our shadows. It has been an honor for me to see them, to have followed them on the dangerous path between something and nothing, between being and extinction. They changed me. They have helped me to believe in the possibility of another life, and a mindlessness that is free, where joy and rage warp into a stupid cry.

I know when I left through the back door. It was about the same time that I made a trip to the mountains of central British Columbia to be with trumpeter swans. Called in Latin *Olor buccinator*, the trumpeter swan has never been on the official list of endangered or threatened species, but they had once become very rare and were a species of deep concern. It was January, and the place I went to was called, appropriately, Lonesome Lake. You could get to it only by about 50 miles of back-country trails or a half-hour flight on a chartered ski-plane. I took the plane. It dropped a friend and me off on the January ice of the lake in the winter dusk of late afternoon, and since we didn't know the area, we pitched our tent in the gathering shadows on the open ice. We slept that night on the lake.

It was a strange night. Only two days earlier, I had been in Rome, after spending a month in Europe. Rome—that ancient and heavy city, full of art, weighted by the architectural monuments of 2000 years. Dwarfed by St. Peter's, I walked the streets, feeling the pressure of all the buildings, erected on top of each other like so many oppressive layers of culture, stones upon stones—classical and Christian and swirling, dizzying baroque. And then I was halfway around the world, in a remote part of British Columbia, suffering from jet lag on a frozen lake, closed in on all sides by dark and porpoising hills. That night, I awoke a couple of times to loud rumbles, like the hooves of moose thundering across the ice. Like gunshot reports of noise. In the cold, the ice shifted, split apart in long, groaning cracks, and the whole lake shivered beneath us in our sleep.

A cultural back door.

The next morning, the sun did not clear the surrounding hills until late. In the bottom of the mountain bowl, we fixed breakfast in the shadows while the light slowly worked its way in a line down the steep hillsides. The rich greens of hemlocks and Douglas firs and stately cedars simmered into burnished yellows. The light finally broke full above the hills and flooded the lake with the sudden cold transparency of the winter sun. Ice crystals spangled on the lake in thousands of dazzling sequins.

In their own flood of white, the trumpeter swans burst upon us at the same time. They had spent the night at the other end of the lake, gathered by open water at an inlet. Huge birds, with 6-foot wingspans, their big bodies and long necks are a satiny, pillowy white, shading to darkness in places, darkness shading to dream. At least 350 swans made up the wintering flock on Lonesome Lake, and they all came at us, skimming only a few feet above the ice, white on glittering white, like waves of planes flying low formations in some massive invasionary force. They are fed in the winter by the Canadian government, and they must have thought we had barley for them.

Soon the swans were swarming all around me and friend and tent in an aggressive sea of white. Restless with premigration energy, they charged and strutted, poked and bit. Their heads were higher than my waist, and they stuck their black beaks into my jacket and jabbed at my pants. The clamor they raised was deafening, a cacophony of honks and screams like bagpipes out of tune. It reminded me of the time I was stranded at the base of a statue in the piazza of Santa Maria del Popolo, in the rush-hour traffic of Rome—cars streaming all around me in a lawless frenzy, horns blasting, all that Italian passion gone mad behind the wheel.

I gave myself over to the swans. I quit thinking, quit trying to understand the chaotic abundance crashing all around me. Here was a chance for a moment of pure stupidness, out of which life can become new again and fresh.

Some of the groups on the ice were families—parents and cygnets. I watched a fight break out on the ice. One hyped-up swan charged into a family group of four. The cob rushed after the intruder in a screaming attack. He chased after the single swan in a fury of white, wings open, neck tense and long. The two of them raced through the milling crowd of assembled swans near us, knocking over bystanders, the enraged cob biting the fleeing swan's neck. Victory ensured, the cob took a painful tug at the tail feathers of the defeated intruder and then swaggered proudly back to his family on pigeon-toed feet.

Reunited, the cob and his mate faced each other and began a ceremony of triumph which looked vaguely like a square dance. Their bodies went all aquiver, their heads bobbed up and down on sinuously lovely necks, wings extended and fluttering, tips down, as if wrapping themselves in white robes. The cygnets, eager to join the dance, imitated their parents with clumsy body shakes, heads bobbing awkwardly.

I loved the dance and found myself shuffling my feet in my own imitation, jumping up and down. Trying to fly? Trying to jump out of myself?

As spectacular as the birds on the ice were the birds in the air. Taking off and landing, swans everywhere: The lake was like a massive concourse at an airport. Small groups made short running starts and lifted off, banking into turns with their necks taut as ropes, wings flapping and flashing in the sun like shattering mirrors. Their incomparably beautiful necks. And light on their wings, heartbreaking in its shimmer and sheen.

Other swans flew in fixed formations up over the hills. In groups of three or four or five, they circled against the pale winter sky and orbited back down to the lake like visions descending.

I reeled with them, overwhelmed by their noise and their energy and the whiteness of ice and sun and swans. Is this, I thought, how Leda felt when the swan appeared to her, overcome by the power of the divine in animal form?

At the other end of the lake, six swans took off and flew toward me in a low glide. I watched them coming the whole way, getting more and more excited. I stayed silent, letting the swans around me make all the noise. But inside, I was all humility and supplication, lost in the anarchy of swans, swept up in what Yeats calls, in "Leda and the Swan," a white rush. Come on, I thought, come on. Take me. Take me. Make me new.

The six swans kept coming, steady, flying barely three feet off the ice, flapping their wings only occasionally. About 10 feet away from me, they pulled up on their wings, which billowed like sails in a wind. Catching the air with their wings, the swans pounded out a booming whoosh and landed on feet that had to run to keep up with the momentum of their big bodies. Those trumpeter swans might as well have crashed into me. They might as well have knocked me over. They might as well have flown right through me, the way a flock of birds will stream into a tree in full leaf and disappear.

APPENDIX 1

Partial List of Extinctions in the United States and U.S. Territories

Table 1 presents a compilation of species and subspecies that have gone extinct in the United States since 1750. There is no "official" list of U.S. extinctions. This table was compiled from the work of Paul Opler, who interviewed U.S. Fish and Wildlife Service specialists on the animals involved; from the Federal Register, January 6, 1989, the Department of the Interior, Fish and Wildlife Service, "Endangered and Threatened Wildlife; Animal Notice of Review"; from a list of extinct vertebrates by James D. Williams and Ronald M. Novak in "Vanishing Species in Our Own Backyard," a chapter in Les Kaufman and Kenneth Mallory, eds., *The Last Extinction*; and from interviews with Michael Bean of the Environmental Defense Fund.

Table 2 lists eleven species that have been identified in a December 21, 1988, Government Accounting Office report, "Endangered Species: Management Improvements Could Enhance Recovery Program." Although these species are still included on the endangered list, they are believed to be extinct. They have yet to be officially declared extinct. Note the presence on this list of the ivory-billed woodpecker. It may be extinct in the United States. Although recently sighted in Cuba, it may soon be extinct there as well. Note also the inclusion of the Bachman's warbler on this list. The dusky seaside sparrow was included in this group by the GAO, but I have transferred it to Table 1.

Table 3 lists three species which have been brought into captivity and are no longer thought to have individuals extant in the wild.

Table 1   Extinctions since 1750

Common Name	Scientific Name	Region	Year*
VERTEBRATES			
FISH			
Chub, Independence Valley Tui	*Gilabicolor isolata*	Nevada	1989 (ext.)
Chub, Thicktail	*Gila crassicauda*	California	1955
Cisco, Blackfin	*Coregonus nigripinnis*	Great Lakes	1960s
Cisco, Deepwater	*Coregonus johannae*	Great Lakes	1951
Cisco, Longjaw	*Coregonus alpenae*	Great Lakes	1970
Dace, Grass Valley Speckled	*Rhinicthys osculus religuus*	Nevada	1938
Dace, Las Vegas	*Rhinichthys deaconi*	Nevada	1950s
Gambusia, Amistad†	*Gambusia amistadensis*	Texas	1977
Killifish, Ash Meadows	*Empetrichthys merriami*	Nevada	1957
Killifish, Pahrump Ranch	*Empetrichthys latos pahrump*	Nevada	1956
Killifish, Raycroft Ranch	*Empetrichthys latos concouus*	Nevada	1956
Lamprey, Miller Lake	*Lampetra minima*	Oregon	1953

*The dates listed indicate the last time the creature was seen in the wild, with these exceptions:

"1989 (ext.)" means that, according to the January 6, 1989, Federal Register, "Endangered and Threatened Wildlife and Plants; Animal Notice of Review," the Fish and Wildlife Service has persuasive evidence of extinction for these taxa. This date was used unless a more specific date was available.

A date in parentheses is the date of death in a zoo.

A dash indicates that the date of last sighting is unknown.

The amistad gambusia was a desert fish, the last known survivors of which were collected for rearing and breeding in captivity. They were accidentally hybridized, and the species was lost.

Table 1   (Continued)

Common Name	Scientific Name	Region	Year
VERTEBRATES			
Madtom, Smoky	*Noturus baileyi*	Tennessee	1957
Minnow, Clear Lake	*Endermichthys grandipennis*	California	1940
Pike, Blue	*Stizostedion vitreum glaucum*	Great Lakes	1971
Pupfish, Monkey Springs	*Cyprinodon sp.*	Arizona	1989 (ext.)
Pupfish, Shoshone	*Cyprinodon nevadensis shoshone*	California	1966
Pupfish, Tecopa	*Cyprinodon nevadensis calidae*	California	1942
Sculpin, Utah Lake	*Cottus echinatus*	Utah	1928
Shiner, Bluntnose	*Notropis simus simus*	New Mexico, Texas	1964
Shiner, Phantom	*Notropis orca*	New Mexico, Texas	1975
Spinedace, Big Spring	*Lepidomeda mollispinus*	Nevada	1950s
Spinedace, Pahrangat	*Lepidomeda altivelis*	Nevada	1950s
Sucker, Harelip	*Lagachila lacera*	Mississippi drainage	1893
Sucker, June	*Chasmistes liorus*	Utah	1935
Sucker, Snake	*Chasmistes muriei*	Wyoming	1928
Topminnow, Whiteline	*Fundulus albolineatus*	Tennessee	1899
Trout, Alvord Cutthroat	*Salmo clarki ssp.*	Illinois, Indiana, Michigan, Wisconsin, Canada	1989 (ext.)
Trout, Silver	*Salvelinus agassizi*	New Hampshire	1930s
Trout, Yellowfin Cutthroat	*Salmo clarki macdonaldi*	Colorado	1910
AMPHIBIANS			
Frog, Las Vegas Leopard	*Rana onca*	Nevada	1938
Frog, San Felipe Leopard	*Rana sp.*	California	1989 (ext.)

Table 1   (Continued)

Common Name	Scientific Name	Region	Year
VERTEBRATES			
REPTILES			
Iguana, Navassa Island	*Cyclura cornuta nigerrima*	West Indies	1989 (ext.)
Lizard, Navassa Island Curley-tailed	*Leiocephalus eremmitus*	West Indies	1800s
Racer, St. Croix (ground snake)	*Alsophis sancticrucis*	Virgin Islands	1900s
BIRDS			
Akepa, Oahu	*Loxops coccinea rufa*	Hawaii	1893
Akialoa, Hawaii	*Hemignathus obscurus obscurus*	Hawaii	1940
Akialoa, Lanai	*Hemignathus obscurus lanaiensis*	Hawaii	1894
Akialoa, Oahu	*Hemignathus obscurus ellisianus*	Hawaii	1837
Amakihi, Greater	*Loxops sagittirostris*	Hawaii	1900
Apapane, Laysan	*Himatione sanguinea freeethi*	Hawaii	1923
Auk, Great	*Pinguinus impennis*	North Atlantic	June 1844
Crake, Kusaie	*Aphanolimnas monasa*	Caroline Islands	1828
Creeper, Lanai	*Loxops maculata montana*	Hawaii	1937
Dove, Guam White-throated Ground	*Gallicolumba xanthonura xanthonura*	Guam, Mariana Islands	1989 (ext.)
Dove, Mariana Fruit	*Ptilinopus roseicapillus*	Guam, Mariana Islands	1989 (ext.)
Duck, Labrador	*Camptorhynchus labradorius*	Atlantic coast	Dec. 12, 1875
Fantail, Guam	*Rhipidura rufifrons*	Guam	1989 (ext.)

Table 1   (Continued)

Common Name	Scientific Name	Region	Year
VERTEBRATES			
Finch, Greater Kona	Psittirostra palmeri	Hawaii	1896
Finch, Kona	Psittirostra kona	Hawaii	1894
Finch, Lesser Kona	Psittirostra flaviceps	Hawaii	1891
Hen, Heath	Tympanuchus cupido cupido	Atlantic coast	Mar. 11, 1932
Honey-eater, Cardinal	Myxomela cardinalis saffordi	Guam, Mariana Islands	1989 (ext.)
Kioea	Chaetoptila angustipluma	Hawaii	1859
Mamo	Drepanis pacifica	Hawaii	1898
Mamo, Black (oo-nuku-umu)	Drepanis funerea	Hawaii	1907
Millerbird, Laysan	Acrocephalus familiaris	Hawaii	1923
Nukupu'u, Oahu	Hemignathus lucidus lucidus	Hawaii	1860
O-o, Hawaii	Moho nobilis	Hawaii	1934
O-o, Molokai	Moho bishopi	Hawaii	1915
O-o, Oahu	Moho apicalis	Hawaii	1837
Owl, Virgin Islands Screech	Otus nudipes newtoni	Puerto Rico, Virgin Islands	1980
Parakeet, Carolina	Conuropsis carolinensis	Southeast	1904 (Zoo, 1914)
Parakeet, Mauge's	Aratinga chloroptera maugei	Puerto Rico	1892
Pigeon, Passenger (last bird named "Martha")	Ectopistes migratorius	Northeast	1899 (Zoo, 1914)
Rail, Hawaiian Brown	Pennula millsi	Hawaii	1964
Rail, Hawaiian Spotted	Pennula sandwichensis	Hawaii	1893
Rail, Laysan	Porzanula palmeri	Hawaii	1944
Rail, Wake Island	Rallus wakensis	Wake Island	1945
Sparrow, Dusky Seaside	Ammospiza maritima nigrescens	Florida	1989 (Zoo, June 16, 1987)

Table 1 (Continued)

Common Name	Scientific Name	Region	Year
VERTEBRATES			
Sparrow, Santa Barbara Song	*Melospiza melodia graminea*	Santa Barbara Islands	1967
Sparrow, Smyrna Seaside‡	*Ammospiza maritima pelonota*	Florida	—
Sparrow, Texas Henslow's	*Ammodramus henslowii houstonensis*	Texas	1983
Starling, Kusaie	*Aplonis corvina*	Caroline Islands	1828
Thrush, Lanai	*Phaeornis obscurus lanaiensis*	Hawaii	1931
Thrush, Oahu	*Phaeornis obscurus oahensis*	Hawaii	1825
Ula-ai-hawane	*Ciridops ana*	Hawaii	1892
Wren, San Clemente Bewick's	*Thryomanes bewickii leucophrys*	California	1927
MAMMALS			
Bat, Puerto Rican Long-nosed	*Monophylus plethodon frater*	Puerto Rico	1900?
Bat, Puerto Rican Long-tongued	*Phyllonycteris major*	Puerto Rico	1900?
Bear, California Grizzly	*Ursus arctos californicus*	California	1925
Bighorn, Badlands	*Ovis canadensis auduboni*	Great Plains	1910
Caribou, Woodland (Montana pop.)	*Rangifer tarandus caribou*	Montana	1989 (ext.)
Chipmunk, Penasco	*Eutamias minimus atristriatus*	New Mexico	1980
Cougar, Wisconsin	*Felis concolor schorgeri*	North central states	1930

‡The Smyrna Seaside Sparrow has not been officially listed, though it was a candidate for listing. It has long been feared extinct. As of 1987, the U.S. Fish and Wildlife Service searched for this sparrow and could not find it. The bird is increasingly believed to be extinct.

Table 1 (Continued)

Common Name	Scientific Name	Region	Year
VERTEBRATES			
Elk, Eastern	Cervus canadensis canadensis	Plains and east	1880
Elk, Merriam's	Cervus canadensis merriami	Arizona	1906
Fox, Southern California Kit	Vulpes macrotis macrotis	California	1903
Gopher, Goff's Pocket	Geomys pinetis goffi	Florida	1955
Gopher, Sherman's Pocket	Geomys pinetis fontanelus	Georgia	1950
Gopher, Tacoma Pocket	Thomomys mazama tacomensis	Washington	1970
Mink, Sea	Mustela macrodon	New England coast	1890
Mouse, Anastasia Island	Peromyscus gossypinus anastasae	Florida	1989 (ext.)
Mouse, Chadwick Beach Cotton	Peromyscus gossypinus restrictus	Florida	1950?
Mouse, Giant Deer	Peromyscus nesodytes	Channel Islands, California	1870
Mouse, Pallid Beach	Peromyscus polionotus decoloratus	Florida	1946
Sea Cow, Steller's	Hydrodamalis stelleri	Aleutian Islands	1768
Seal, Caribbean Monk	Monachus tropicalis	Florida Gulf	1962
Vole, Amargosa Meadow	Microtus californicus scripensis	California	1917
Vole, Gull Island	Microtus nesophilus	New York	1898
Vole, Louisiana	Microtus orchrogaster ludovicianus	Louisiana, Texas	1905
Whale, Atlantic Gray	Eschrichtius gibbosus gibbosus	Atlantic coast	1750

Table 1   (Continued)

Common Name	Scientific Name	Region	Year
VERTEBRATES			
Wolf, Cascade Mountains	*Canis lupus fuscus*	British Columbia, Oregon, Washington	1940
Wolf, Florida Red	*Canis rufus floridanus*	Southeast	1930
Wolf, Kenai	*Canis lupus alces*	Alaska	1925
Wolf, Mogollon Mountain	*Canis lupus mogollonensis*	Southwest	1935
Wolf, Plains	*Canis lupus nubilus*	Great Plains	1926
Wolf, Southern Rocky Mountain	*Canis lupus youngi*	Rocky Mountains	1935
Wolf, Texas Gray	*Canis lupus monstrabilis*	Texas	1942
Wolf, Texas Red	*Canis rufus rufus*	Oklahoma, Texas	1970
INVERTEBRATES			
FLATWORMS			
Planarian, Bigger's Groundwater	*Sphalloplana subtilis*	Virginia	1989 (ext.)
Planarian, Holsinger's Groundwater	*Sphalloplana holsingeri*	Virginia	1989 (ext.)
CRUSTACEANS			
Amphipod, Rubious Cave	*Stygobromus lucifugus*	Illinois	1989 (ext.)
Crayfish, Sooty	*Pacifastacus nigrescens*	California	1860s
Scud, Hay's Spring	*Stygonectes hayi*	District of Columbia	1957
Shrimp, Pasadena	*Syncaris pasadenae*	California	1933
INSECTS (Mayflies through Flies)			
Beetle, Duck River	*Pseudanophthalmus tullahoma*	Tennessee	1989 (ext.)
Beetle, Mono Lake Hygrotus Diving	*Hygrotus artus*	California	1989 (ext.)
Beetle, Nickajack Cave	*Pseudanophthalmus nickajackensis*	Virginia	1989 (ext.)

Table 1 (Continued)

Common Name	Scientific Name	Region	Year
INVERTEBRATES			
Beetle, Tooth Cave Blind Rove	*Cylindropsis sp.*	Texas	1989 (ext.)
Bug, Phyllostegian Leaf	*Cytropeltis phyllostegiae*	Hawaii	1989 (ext.)
Damselfly, Jugorum Megalagrion	*Megalagrion jugorum*	Hawaii	1989 (ext.)
Damselfly, Nesiotes Megalagrion	*Megalagrion nesiotes*	Hawaii	1989 (ext.)
Fly, Hawaiian Chersodromian Dance	*Chersodromia hawaiiensis*	Hawaii	1989 (ext.)
Fly, Ko'olau Spurwing Long-legged	*Campsicnemus mirabilis*	Hawaii	1989 (ext.)
Fly, Lanai Pomace	*Drosophila lanaiensis*	Hawaii	1989 (ext.)
Fly, Valley Mydas	*Raphiomydas trochilus*	California	1989 (ext.)
Fly, Volutine Stonemyian Tabanid	*Stonemyia volutina*	California	1989 (ext.)
Katydid, Antioch Dunes Shieldback	*Neduba extincta*	California	1989 (ext.)
Katydid, Remote Conehead	*Conocephaloides remotus*	Hawaii	1989 (ext.)
Mayfly, Diverse Isonychian	*Isonychia diversa*	Tennessee	1989 (ext.)
Mayfly, Robust Pentagenian Burrowing	*Pentagenia robusta*	Ohio	1989 (ext.)
Scale	*Clavicoccus erinaceus*	Hawaii	—
Scale	*Phylococcus oahuensis*	Hawaii	—
Stonefly, Robert's Alloperlan	*Alloperla roberti*	Illinois	1989 (ext.)
Weevil	*Dryophthorus distinguendus*	Hawaii	—
Weevil	*Dryotribus mimeticus*	Hawaii	—

Table 1 (Continued)

Common Name	Scientific Name	Region	Year
INVERTEBRATES			
Weevil	*Macrancylus linearis*	Hawaii	—
Weevil	*Oedames laysanensis*	Hawaii	—
Weevil	*Pentarthrum blackburni*	Hawaii	—
Weevil, Fort Ross Trigonoscuta	*Trigonoscuta rossi*	California	1989 (ext.)
Weevil, Yorba Linda Trigonoscuta	*Trigonoscuta yorbalindae*	California	1989 (ext.)
INSECTS (Butterflies and Moths)			
Butterfly, Atossa Fritillary	*Speyeria adiaste atossa*	California	1959
Butterfly, Fender's Blue	*Icaricia icarioides fenderi*	Oregon	1989 (ext.)
Butterfly, Silverspot	*Speyeria hydaspe conquista*	New Mexico	1932
Butterfly, Sthenele Wood Nymph	*Cercyonis sthenele sthenele*	California	1890s
Butterfly, Strohbeen's Parnassian	*Parnassius clodius strohbeeni*	California	1958
Butterfly, Texas Tailed Blue	*Everes comyntas texanus*	Texas	1926
Butterfly, Willamette Silverspot	*Speyeria callippe extincta*	Oregon	1989 (ext.)
Butterfly, Xerces Blue	*Glaucopsyche xerces*	California	1943
Moth, "Ola" (a peppered looper)	*Tritocleis microphylla*	Hawaii	1989 (ext.)
Moth, "Poko" Noctuid	*Agrotis crinigera*	Hawaii	1989 (ext.)
Moth, American Chestnut Nepticulid	*Ectodemia castaneae*	Maryland	1989 (ext.)

Table 1 (Continued)

Common Name	Scientific Name	Region	Year
INVERTEBRATES			
Moth, Blackburn's Sphinx	*Manduca blackburni*	Hawaii	1989 (ext.)
Moth, Chestnut Casebearer	*Coleophora leucochrysella*	Pennsylvania	1989 (ext.)
Moth, Chestnut Ermine	*Argyresthia castaneela*	New Hampshire, Vermont	1989 (ext.)
Moth, Chestnut Leaf Miner	*Tischeria perplexa*	Virginia	1989 (ext.)
Moth, Confused Helicoverpan Noctuid	*Helicoverpa confusa*	Hawaii	1989 (ext.)
Moth, Hawaiian Hopseed Looper	*Scotorythra paratactis*	Hawaii	1989 (ext.)
Moth, Hilo Hypenan Noctuid	*Hypena newelli*	Hawaii	1989 (ext.)
Moth, Kaholuamano Noctuid	*Hypena senicula*	Hawaii	1989 (ext.)
Moth, Kerr's Agrotis Noctuid	*Agrotis kerri*	Hawaii	1989 (ext.)
Moth, Ko'olau Giant Looper	*Scotorythra nesiotes*	Hawaii	1989 (ext.)
Moth, Kona Giant Looper	*Scotorythra megalophylla*	Hawaii	1989 (ext.)
Moth, Laysan Agrotis Noctuid	*Agrotis laysanensis*	Hawaii	1989 (ext.)
Moth, Laysan Dropseed Noctuid	*Hypena laysanensis*	Hawaii	1989 (ext.)
Moth, Laysan Hedyleptan	*Hedylepta laysanensis*	Hawaii	1989 (ext.)
Moth, Lovegrass Noctuid	*Hypena plagiota*	Hawaii	1989 (ext.)
Moth, Midway Agrotis Noctuid	*Agrotis fasciata*	Hawaii	1989 (ext.)
Moth, Minute Helicoverpan Noctuid	*Helicoverpa minuta*	Hawaii	1989 (ext.)
Moth, Noctuid	*Peridroma porphyrea*	Hawaii	—

Table 1   (Continued)

Common Name	Scientific Name	Region	Year
INVERTEBRATES			
Moth, Oahu Swamp Hedyleptan	*Hedylepta epicentra*	Hawaii	1989 (ext.)
Moth, Phleophagan Chestnut Nepticulid	*Ectodemia phleophaga*	Maryland	1989 (ext.)
Moth, Procellaris Agrotis Noctuid	*Agrotis procellaris*	Hawaii	1989 (ext.)
Moth, Telegraphic Hedyleptan	*Hedylepta telegrapha*	Hawaii	1989 (ext.)

INSECTS (Caddis flies and Bees)

Common Name	Scientific Name	Region	Year
Bee, Blackburn's Yellow-faced	*Nesoprosopis blackburni*	Hawaii	1989 (ext.)
Bee, Bristlefront Yellow-faced	*Nesoprosopis setosifrons*	Hawaii	1989 (ext.)
Bee, Broadhead Yellow-faced	*Nesoprosopis laticeps*	Hawaii	1989 (ext.)
Bee, Connected Yellow-faced	*Nesoprosopis connectens*	Hawaii	1989 (ext.)
Bee, Erythrodeme Yellow-faced	*Nesoprosopis erythrodemas*	Hawaii	1989 (ext.)
Bee, Finitiman Yellow-faced	*Nesoprosopis finitima*	Hawaii	1989 (ext.)
Bee, Hilaris Yellow-faced	*Nesoprosopis hilaris*	Hawaii	1989 (ext.)
Bee, Lanai Yellow-faced	*Nesoprosopis angustula*	Hawaii	1989 (ext.)
Bee, Maui Yellow-faced	*Nesoprosopis mauiensis*	Hawaii	1989 (ext.)
Bee, Melanothrix Yellow-faced	*Nesoprosopis melanothrix*	Hawaii	1989 (ext.)
Bee, Molokai Yellow-faced	*Nesoprosopis neglecta*	Hawaii	1989 (ext.)
Bee, Monocolor Yellow-faced	*Nesoprosopis homeochroma*	Hawaii	1989 (ext.)
Bee, Mutatan Yellow-faced	*Nesoprosopis mutata*	Hawaii	1989 (ext.)
Bee, Pele Yellow-faced	*Nesoprosopis pele*	Hawaii	1989 (ext.)

Table 1   (Continued)

Common Name	Scientific Name	Region	Year
	INVERTEBRATES		
Bee, Perspicuan Yellow-faced	*Nesoprosopis perspicua*	Hawaii	1989 (ext.)
Bee, Psammobian Yellow-faced	*Nesoprosopis psammobia*	Hawaii	1989 (ext.)
Bee, Rugulose Yellow-faced	*Nesoprosopis rugulosa*	Hawaii	1989 (ext.)
Bee, Snowy Yellow-faced	*Nesoprosopis nivalis*	Hawaii	1989 (ext.)
Caddisfly, Athens Long-horned	*Triaenodes phalacris*	Ohio	1989 (ext.)
SNAILS			
Magnificent, Alabama	*Tulatoma magnifica*	Alabama	1957
Magnificent, North Carolina	*Helisoma magnifica*	North Carolina	1947
Magnificent, Texas	*Taphius encosmius*	Texas	1947
Marshsnail, Fish Springs	*Stagnicola pilsbryi*	Utah	1989 (ext.)
Snail, Alabama Live Bearing	*Tulatoma angulata*	Alabama	1957
Snail, Alabama River	*Apella alabamensis*	Alabama	1924
Snail, Alabama River	*Apella amplum*	Alabama	1924
Snail, Alabama River	*Apella cariniferum*	Alabama	1924
Snail, Alabama River	*Apella excisum*	Alabama	1924
Snail, Alabama River	*Apella hendersoni*	Alabama	1924
Snail, Alabama River	*Apella incisum*	Alabama	1924
Snail, Alabama River	*Apella laciniatum*	Alabama	1924
Snail, Alabama River	*Apella lewisi*	Alabama	1924
Snail, Alabama River	*Apella pagodum*	Alabama	1924
Snail, Alabama River	*Apella pumilum*	Alabama	1924

Table 1   (Continued)

Common Name	Scientific Name	Region	Year
INVERTEBRATES			
Snail, Alabama River	*Apella pyramidatum*	Alabama	1924
Snail, Alabama River	*Apella spillmanii*	Alabama	1924
Snail, Alabama River	*Apella walkeri*	Alabama	1924
Snail, Avalon	*Oreohelix avalonensis*	California	1891
Snail, Catenoid River	*Oxytrema catenoides*	Georgia	1800s
Snail, Coosa Live Bearing	*Tulatoma coosaensis*	Alabama	1957
Snail, Hawaiian Land	*Carelia anceophila*	Hawaii	1931
Snail, Hawaiian Land	*Carelia cumingiana*	Hawaii	1931
Snail, Hawaiian Land	*Carelia glossema*	Hawaii	1931
Snail, Hawaiian Land	*Carelia kalalauensis*	Hawaii	1931
Snail, Hawaiian Land	*Carelia knudseni*	Hawaii	1931
Snail, Hawaiian Land	*Carelia olivacea*	Hawaii	1931
Snail, Hawaiian Land	*Carelia paradoxa*	Hawaii	1931
Snail, Hawaiian Land	*Carelia perscelis*	Hawaii	1931
Snail, Hawaiian Land	*Carelia tenebrosa*	Hawaii	1931
Snail, Hawaiian Land	*Carelia turricula*	Hawaii	1931
Snail, Hawaiian Tree	*Achatinella abbreviala*	Hawaii	—
Snail, Hawaiian Tree	*Achatinella buddi*	Hawaii	—
Snail, Hawaiian Tree	*Achatinella byroni*	Hawaii	—
Snail, Hawaiian Tree	*Achatinella caesia*	Hawaii	—

Table 1 (Continued)

Common Name	Scientific Name	Region	Year
		INVERTEBRATES	
Snail, Hawaiian Tree	*Achatinella casta*	Hawaii	—
Snail, Hawaiian Tree	*Achatinella decora*	Hawaii	—
Snail, Hawaiian Tree	*Achatinella juncea*	Hawaii	—
Snail, Hawaiian Tree	*Achatinella lehuiensis*	Hawaii	—
Snail, Hawaiian Tree	*Achatinella lila*	Hawaii	—
Snail, Hawaiian Tree	*Achatinella payracea*	Hawaii	—
Snail, Hawaiian Tree	*Achatinella rosea*	Hawaii	—
Snail, Hawaiian Tree	*Achatinella spaldingi*	Hawaii	—
Snail, Hawaiian Tree	*Achatinella thaanumi*	Hawaii	—
Snail, Hawaiian Tree	*Achatinella turgida*	Hawaii	—
Snail, Longstreet Spring	*"Fluminicola" sp.*	Nevada	1989 (ext.)
Snail, Socorro	*Amnicola neomexicana*	New Mexico	1971
MUSSELS			
Acornshell (formerly Acorn Pearly Mussel)	*Epiblasma haysiana*	Alabama, Tennessee, Virginia	1989 (ext.)
Catspaw, Narrow (formerly Stones Pearly Mussel)	*Epioblasma lenoir*	Alabama, Tennessee	1965
Elktoe, Carolina (Mussel)	*Alasmidonta robusta*	North Carolina	1989 (ext.)
Elktoe, Coosa (Mussel)	*Alasmidonta maccordi*	Alabama	1989 (ext.)
Forkshell (formerly Lewis Pearly Mussel)	*Epioblasma lewisi*	Alabama, Tennessee, Kentucky	1940s

Table 1 (Continued)

Common Name	Scientific Name	Region	Year
INVERTEBRATES			
Leafshell (formerly Arcuate Pearly Mussel)	*Epioblasma flexuosa*	Alabama, Tennessee	1940s
Leafshell, Cumberland (formerly Steward's Pearly Mussel)	*Epioblasma stewardsoni*	Alabama, Tennessee	1910s
Mussel, Recovery Pearly	*Elliptio nigella*	Georgia	1954
Mussel, Sampson's Pearly	*Epioblasma sampsoni*	Indiana, Illinois	1910
Riffleshell, Angled	*Epioblasma biemarginata*	Alabama, Tennessee	1989 (ext.)
Riffleshell, Tennessee (formerly Nearby Pearly Mussel)	*Epioblasma propinqua*	Alabama, Tennessee	1910s
Round combshell (formerly Fine-rayed Pearly Mussel)	*Epioblasma personata*	Ohio River	1800s
Sugarspoon (formerly Arc-form Pearly Mussel)	*Epioblasma arcaeformis*	Alabama, Tennessee	1940s

Table 2   Conjectured Extinctions

Common Name	Scientific Name	Region
Anole, Giant (lizard)	*Anolis roosevelti*	Puerto Rico
Bat, Mariana Fruit	*Pteropus marianus*	Mariana Islands
Butterfly, Palos Verdes Blue*	*Glaucopsyche lygdamus palosveresensis*	California
Cougar, Eastern	*Felis concolor couguar*	Eastern U.S.
Madtom, Scioto (fish)	*Noturus trautmani*	Ohio
Mallard, Mariana	*Anas oestaleti mariana*	Mariana Islands
Mussel, Tubercled-blossom Pearly	*Epioblasma torulosa torulosa*	Indiana, Illinois, Kentucky, Tennessee
Mussel, Turgid-blossom Pearly	*Epioblasma turgidula*	Alabama, Tennessee
Mussel, Yellow-blossom Pearly	*Epioblasma florentina florentina*	Alabama, Tennessee
Warbler, Bachman's	*Vermivora bachmanii*	Southeast
Woodpecker, Ivory-billed	*Campephilus principalis*	Southeast

*The Palos Verdes Blue Butterfly probably went extinct in the early 1980s when its habitat in Los Angeles was destroyed for a park—Rancho Palos Verdes.

Table 3   Extinctions in the Wild

Common Name	Scientific Name	Final Captures
Black-footed Ferret	*Mustela nigripes*	Feb. 28, 1987
California Condor	*Gymnogyps californicus*	Apr. 19, 1987
Guam Rail	*Rallus owstoni*	Mar. 1, 1985

APPENDIX 2

Organizations Concerned with Wildlife, Extinctions, and the Environment

You may now feel the need to do something to help endangered wildlife. Perhaps the most important step anyone can take is to examine his or her heart, and to change his or her life. We need to become more aware of the implications of our daily lives and, it seems to me, learn to see how many of our realities are also deeply rooted in attitudes.

In addition to exploring such personal matters, you can join any of a number of organizations throughout the United States that confront the issues involved in protecting the animals. Most of them rely heavily upon volunteers, and you can become as active as your time and talents and conscience dictate. The activities of these organizations, and the nature of their politics, vary widely—from conservative to radical, from contemplative to active. Some try to remain apolitical. Many others have grown increasingly political, as they have realized that nature study—once an end in itself—can no longer be conducted in a social vacuum. Without active involvement in politics, without protection, there might quickly be little nature left to study.

The list below gives a sampling of organizations with different emphases; their inclusion on this list is less an endorsement than an invitation for you to learn more for yourself. They derive largely from the *Conservation Directory*, published annually by the National Wildlife Federation. You can be sure that any interest you have, even if not reflected in the list below, is likely to be represented by an organization somewhere, and I highly recommend that you consult the *Conservation Directory*. In it, you will find groups organized to support individual species as well as groups organized to protect local or regional environmental concerns. While I have mainly stressed organizations focusing on the United States and Canada, many others are concerned with international conservation efforts. In preparing this list, I have also drawn upon the list of organizations that appears at the end of *The Last Extinction*, edited by Les Kaufman and Kenneth Mallory.

Alliance for Environmental Education, Inc.
Box 1040
3421 M Street, NW
Washington, DC 20007

A consortium of 35 organizations whose goal is to further educational activities that enhance the personal and social quality of life.

American Association of Zoological Parks and Aquariums
Oglebay Park
Wheeling, WV 26003

Vital programs include conservation of wildlife and preservation and propagation of rare and endangered species.

American Cetacean Society
P.O. Box 2639
San Pedro, CA 90731

Threefold focus on conservation, education, and involvement with all aquatic matters, especially whales, dolphins, and porpoises.

American Hiking Society
1015 31st Street, NW
Washington, DC 20007

Focuses on the education of the public concerning hiking and on the protection of the interests of hikers.

American Land Resource Association
1516 P Street, NW
Washington, DC 20033

Undertakes field studies and disseminates scholarly and popular publications on rural land conservation, renewable resources, and protection of outstanding landscapes and critical land forms.

American Littoral Society
Sandy Hook
Highlands, NJ 07732

Interested in the study and conservation of coastal habitat, barrier beaches, wetlands, estuaries, and near-shore waters, as well as their wildlife.

American Museum of Natural History
Central Park West at 79th Street
New York, NY 10024

Conducts research and educates the public in anthropology and the natural sciences, including living and extinct animals, ecological relationships, and the development of human cultures.

American Rivers
801 Pennsylvania Avenue, SE
Suite 303
Washington, DC 20003

Works on legislation aimed at the protection of wild and scenic rivers at the federal and state levels.

American Society for Environmental History
Department of History
Oregon State University
Corvallis, OR 97331

Seeks understanding of human ecology through the perspectives of history and the humanities.

American Wilderness Alliance
7600 East Arapahoe Road
Suite 114
Englewood, CO 80112

Dedicated to the conservation of the nation's wilderness, wildlife habitat, and wild rivers.

Animal Protection Institute of America
P.O. Box 22505
6130 Freeport Boulevard
Sacramento, CA 95822

Strives to eliminate fear, pain, and suffering among all animals—domestic and wild.

Animal Welfare Institute
P.O. Box 3650
Washington, DC 20007

Works to improve conditions for laboratory animals and to protect endangered species.

Atlantic Center for the Environment
39 South Main Street
Ipswich, MA 01938

Conducts conservation programs through an intern work force, recruited from colleges and universities across North America.

Center for Environmental Education, Inc.
1725 DeSales Street, NW
Suite 500
Washington, DC 20036

Dedicated to conservation of endangered and threatened species and their marine habitats.

Center for Plant Conservation, Inc.
125 Arborway
Jamaica Plain, MA 02130

Serves as a network of botanical gardens and arboreta dedicated to the preservation and study of endangered U.S. plants.

Children of the Green Earth
Box 95219
Seattle, WA 98145

A worldwide association of people who help children plant trees and do other earth-healing work.

Coastal Conservation Association, Inc.
4801 Woodway
Suite 220 West
Houston, TX 77056

Promotes the preservation and protection of the animal and plant life both onshore and offshore along the coastal areas of the U.S.

Conservation Foundation
1250 24th Street, NW
Washington, DC 20037

Conducts research and educational programs in land use, toxic substances, water resources, and air pollution control.

Cousteau Society, Inc.
930 West 21st Street
Norfolk, VA 23517

Dedicated to the protection and improvement of the quality of life.

Defenders of Wildlife
1244 19th Street, NW
Washington, DC 20036

Strives to preserve, enhance, and protect the natural abundance and diversity of wildlife through education and reasoned advocacy of appropriate public policy.

Desert Fishes Council
407 West Line Street
Bishop, CA 93514

Provides for the exchange of information on the status, protection, and management of endemic fauna and flora of North American desert systems.

EarthFirst
P.O. Box 2358
Lewiston, ME 04241

Takes direct action, including "monkey-wrenching," to gain public attention for environmental concerns.

Earthscan
1717 Massachusetts Avenue, NW
Suite 302
Washington, DC 20036

Provides an information service on global resource, environmental, and development issues.

Earthwatch
319 Arlington Street
Watertown, MA 02172

Invites individuals to join scientific expeditions and thereby help defray their costs; emphasizes endangered cultures, animals, and habitats.

Environmental Action, Inc.
1525 New Hampshire Avenue, NW
Washington, DC 20036

Focuses on political and social change covering environmental issues such as nuclear power, acid rain, and others.

Environmental Defense Fund, Inc.
257 Park Avenue South
New York, NY 10010

Organization of lawyers, scientists, and economists dedicated to protecting and improving environmental quality and public health.

Environmental Policy Institute
218 D Street, SE
Washington, DC 20003

Advocacy organization dedicated to influencing public policies affecting natural resources.

Friends of the Earth
530 Seventh Street SE
Washington, DC 20003

Committed to the preservation, restoration, and rational use of the earth.

Fund for Animals, Inc.
200 West 57th Street
New York, NY 10019

Works to preserve wildlife, save endangered species, and promote
humane treatment for all animals.

Global Tomorrow Coalition, Inc.
1325 G Street, NW
Suite 915
Washington, DC 20005

Fosters public understanding in the United States of the long-term
significance of global trends in population, resources, environment, and
development.

Greenpeace, USA, Inc.
1611 Connecticut Avenue, NW
Washington, DC 20009

Dedicated to the preservation of marine ecosystems; uses nonviolent
direct action to confront environmental abuse.

International Council for Bird Preservation
219C Huntingdon Road
Cambridge CB3 ODL
England

Determines status of bird species, compiles data on all endangered
species, identifies problems and priorities, and promotes conservation projects and conventions.

International Ecology Society (IES)
1471 Barclay St.
St. Paul, MN 55106

Dedicated to the protection of the environment and to a better understanding of all life forms.

International Fund for Animal Welfare
P.O. Box 193
Yarmouth Port, MA 02675

Dedicated to the protection of wild and domestic animals.

International Union for Conservation of Nature and Natural
    Resources (IUCN)
Avenue du Mont-Blanc
CH-1196 Gland
Switzerland

Promotes scientifically based action for the conservation of wild living
resources.

Izaak Walton League of America, Inc.
1701 North Fort Myer Drive
Suite 1100
Arlington, VA 22209

Promotes means and opportunities for educating the public to conserve
and restore the natural environment.

Monitor
1506 19th Street, NW
Washington, DC 20036

Consortium that serves as a coordinating center and information clearinghouse on endangered species and marine mammals.

National Audubon Society
950 Third Avenue
New York, NY 10022

Carries out balanced program of research, education, and action for
the protection of land, water, and wildlife.

National Geographic Society
17th and M Streets, NW
Washington, DC 20036

Supports the increase and diffusion of geographic knowledge, which
includes the study of endangered wildlife.

National Wildlife Federation
1412 Sixteenth Street, NW
Washington, DC 20036

Dedicated to creating and encouraging an awareness of the need for
proper management of the resources of the earth.

Natural Resources Defense Council, Inc.
122 East 42d Street
New York, NY 10168

Dedicated to protecting America's endangered natural resources and to improving the quality of the human environment; uses interdisciplinary legal and scientific approach in monitoring government agencies, bringing legal action, and disseminating citizen information.

Nature Conservancy
1800 North Kent Street
Suite 800
Arlington, VA 22209

Committed to preserving biological diversity by protecting natural lands and the life they harbor.

New York Zoological Society
The Zoological Park
Bronx, NY 10460

Promotes zoological research, public understanding of zoology and the environment, and wildlife conservation through publications, educational programs, and the establishment of zoological parks.

North American Association for Environmental Education
P.O. Box 400
Troy, OH 45373

Assists and supports the work of individuals and groups engaged in environmental education, research, and service.

North American Wolf Society
6461 Troy Pike
Versailles, KY 40383

Dedicated to the wise stewardship of the wolf and other wild canids in North America; reports to members on reintroduction potential, scientific research, and available literature and art.

Oceanic Society
P.O. Box 6032 NW
Washington, DC 20005

Works to protect and preserve the marine environment for people and wildlife, using marine policy analysis and individual involvement opportunities around the world.

Pacific Seabird Group
Box 321
Bolinas, CA 94924

Promotes the study and conservation of Pacific seabirds.

Planned Parenthood Federation of America, Inc.
810 Seventh Avenue
New York, NY 10019

Provides family planning services and information to Americans and people in the developing world.

Population-Environment Balance, Inc.
1325 G Street, NW
Suite 1003
Washington, DC 20005

Champions a national commitment to a stable population, strong economy, and an ecologically diverse environment.

Population Reference Bureau, Inc.
2213 M Street, NW
Washington, DC 20037

Gathers, analyzes, and publishes information on the social and environmental implications of U.S. and international population dynamics.

Rachel Carson Council, Inc.
8940 Jones Mill Rd.
Chevy Chase, MD 20815

Acts as clearinghouse for information on ecology, especially concerning chemical contamination.

Rainforest Action Network
466 Green Street
Suite 300
San Francisco, CA 94133

Works to protect the world's tropical rainforests.

Sea Shepherd Conservation Society
P.O. Box 7000-S
Redondo Beach, CA 90277

Pursues field campaigns and educational campaigns directed toward the protection and conservation of marine wildlife.

Sierra Club
730 Polk Street
San Francisco, CA 94109

Promotes the exploration, enjoyment, and protection of wild places on earth, using lawful means to do so.

Smithsonian Institution
1000 Jefferson Drive, SW
Washington, DC 20560

Promotes the "increase and diffusion of knowledge among men" through field investigations, development of national collections in natural history and anthropology, scientific research and publications, and programs that include conservation.

Whale Center
3929 Piedmont Avenue
Oakland, CA 94611

Works for whales and their ocean habitat through conservation, education, research, and advocacy.

Wilderness Society
1400 I Street, NW
10th Floor
Washington, DC 20005

Devoted to preserving wilderness and wildlife, protecting America's prime forests, parks, rivers and shorelines, and fostering an American land ethic.

Wilderness Watch
P.O. Box 782
Sturgeon Bay, WI 54235

Dedicated to sustained use of America's lands and waters, placing ecological considerations foremost.

Wildfowl Foundation, Inc.
2111 Jefferson Davis Highway
605-S
Arlington, VA 22202

Dedicated to advancing the conservation of ducks, geese, and swans of the world by international cooperation in scientific research and education.

Wildlife Information Center, Inc.
629 Green Street
Allentown, PA 18102

Secures and disseminates wildlife conservation, educational, recreational, and scientific information.

World Society for the Protection of Animals
29 Perkins Street
P.O. Box 190
Boston, MA 02130

Promotes the conservation and protection of animals both domestic and wild; formerly, International Society for the Protection of Animals (IPSA).

World Wildlife Fund—U.S.
1250 24th Street, NW
Washington, DC 20037

Works worldwide to protect endangered wildlife and wildlands, especially in the tropical forests of Latin America, Asia, and Africa; supports scientific investigations, monitors international trade in wildlife, promotes ecologically sound development, and assists local groups to take the lead in conservation projects.

Worldwatch Institute
1776 Massachusetts Avenue, NW
Washington, DC 20036

Identifies and analyzes emerging global problems and trends, and brings them to the attention of opinion leaders and the public.

Xerces Society
10 Southwest Ash Street
Portland, OR 97204

Fosters habitat protection for rare and endangered invertebrates, especially butterflies, and seeks to enhance the public's feelings for insects by emphasizing their beneficial roles in ecosystems.

Zero Population Growth, Inc.
1601 Connecticut Avenue, NW
Washington, DC 20009

Works to achieve a balance among people, resources, and environment by advocating stable population worldwide.

SELECT BIBLIOGRAPHY

The writings included in this bibliography have been chosen on the basis of either their usefulness to me in writing this book or their value for a reader who wishes to pursue the issues I have raised. I have not tried to be comprehensive or exhaustive in the choice of these works. Instead, they represent a cross between a trail and a map—indications of where I have been, and suggestions for the reader who wants to do more exploring. I have consulted, and recommend, a wide range of readings, spanning the general phenomenon of endangered species, the biology of particular species, the philosophical issues raised by the loss and preservation of species, and fundamental philosophical questions concerned with competing theories of nature. I also include works—both historical and contemporary—that help in understanding or interpreting endangered species as symbols.

Introduction

For the most thorough and comprehensive overview of the problem of endangered species and extinction, I recommend Paul and Anne Ehrlich, *Extinction: The Causes and Consequences of the Disappearance of Species* (New York, 1981). Also preeminent in the field is Norman Myers. His book *The Sinking Ark: A New Look at the Problem of Disappearing Species* (Oxford, 1983) offers not only documentation and estimates on the rate of the loss of species but also practical suggestions for arresting the trend. See also, by Myers, *A Wealth of Wild Species: Storehouse for Human Survival* (Boulder, Colo., 1983) and *The Primary Source: Tropical Forests and Our Future* (New York, 1985) for an authoritative account of the threat to tropical forests, where the most rapid devastation of nature in the world is occurring. Older, but still excellent, is the pioneering work by David Ehrenfeld, *Conserving Life on Earth* (New York, 1972). A more recent book that is both comprehensive and written for the general public is Les Kaufman and Kenneth Mallory, eds., *The Last Extinction* (Cambridge, Mass., 1986). Another valuable study is Tim Halliday, *Vanishing Birds: Their Natural History and Conservation* (New York, 1978).

All these works document the role of humans in causing the plight of modern endangered animals.

A dated bibliography that is especially useful for material on particular species is Don A. Wood, *Endangered Species: A Bibliography on the World's Rare, Endangered and Recently Extinct Wildlife and Plants* (Stillwater, Okla., Oklahoma State University Environmental Series, no. 3, 1978).

For a succinct article that reviews many of the animals that have gone extinct in the United States and clearly explains some of the biology of endangerment and extinction, see Paul A. Opler, "The Parade of Passing Species: A Survey of Extinctions in the U.S.," *Science Teacher* 44 (December 1976): 30–34. See also Peter Raven, "The Destruction of the Tropics," *Frontiers* 40 (1976): 22–23.

Also important is the Government Accounting Office report, "Endangered Species: Management Improvements Could Enhance Recovery Program," (Washington, D.C., 1988).

Many popular articles have attempted an overview on the phenomenon of endangered species. One article offers a cogent analysis of the status of endangered species in the United States and evaluates the political efforts made so far on their behalf: Roger Di Silvestro, "Our Looming Failure on Endangered Species," *Defenders* 59 (July–August 1984): 20–29. See also the August 1988 issue (vol. 5) of *Endangered Species Update* for a number of excellent articles on endangered species conservation. The issue is dedicated to the topic "A Fifteen Year Retrospective on the Endangered Species Act."

The National Audubon Society now sponsors a series, published annually, that offers articles on individual species as well as essays on the conservation and management of wildlife. See the *Audubon Wildlife Report* (Orlando, Fl., 1985–).

All of the above works offer varying estimates of the scope of the endangered species epidemic and project trends into the future. They also offer bibliographies.

Two essential periodical sources for current information on endangered species and their management are the *Endangered Species Technical Bulletin*, Department of the Interior, U.S. Fish and Wildlife Service, Endangered Species Program, 1976–, and the *Endangered Species Update* (Ann Arbor, Mich., School of Natural Resources, University of Michigan, 1983–). The latter source includes a reprint of the *Endangered Species Technical Bulletin*.

Two government studies also document the problems faced by wildlife; both are prepared by the Council on Environmental Quality. See *Wildlife and America* (Washington, D.C., 1978) and *Global 2000 Report to the President: Entering the Twenty-First Century* (Washington, D.C., 1980).

Also crucial is *Red Data Book*, published by the International Union for Conservation of Nature and Natural Resources, Survival Service Commission. (Morges, Switzerland, 1971 and sequels). It should be available in any good library with scientific holdings.

A number of books focus more specifically on accounts of particular endangered species, giving their natural history and reviewing their decline, or on the historical loss of wildlife in America. The most thorough historical study, with abundant references to contemporary sources, is Peter Matthiessen, *Wildlife in America* (New York, Viking, 1959; revised and reprinted by Viking Penguin, 1987).

Useful earlier studies are William T. Hornaday, *Our Vanishing Wildlife: Its Extermination and Preservation* (New York, 1913); Glover M. Allen, "Extinct and Vanishing Mammals of the Western Hemisphere, with the Marine Species of All the Oceans" (Lancaster, Penn., Special Publication of the American Committee for International Wild Life Protection 11, 1942); and James C. Greenway, Jr., *Extinct and Vanishing Birds of the World* (New York, American Committee for International Wild Life Protection, 1958; revised and reprinted by Dover, 1967). Though global in scope, this last work offers both accounts and analysis of endangered American wildlife.

Among the recent works that recount the stories of many of the most prominent of our endangered species, illustrating the causes of the decline and the efforts at recovery, are John P. S. Mackenzie, *Birds in Peril: A Guide to the Endangered Birds of the United States and Canada* (Boston, 1977), and Charles Cadieux, *These Are the Endangered* (Washington, D.C., 1981). See also Ronald M. Nowak, *Our Endangered Wildlife. Then and Now* (Washington, D.C., 1982).

Two bibliographic sources to check, as well, for sources on endangered wildlife are *Wildlife Index* and *Biological Abstracts.*

Several works deal with the value of animals, or philosophical rationales for preserving species. Most fundamental and influential, though not directly related to the topic of endangered species, is Aldo Leopold, *A Sand County Almanac, and Sketches Here and There* (Oxford, 1949). Focusing on the human need for wildlife and the human bond with other creatures, though from different perspectives, are Mary Midgley, *Animals and Why They Matter* (Athens, Ga., 1984), and Edward O. Wilson, *Biophilia* (Cambridge, Mass., 1984). For a collection of essays on the philosophy of human responsibility for nature, see Bryan G. Norton, ed., *The Preservation of Species: The Value of Biological Diversity* (Princeton, N.J., 1986).

I have found several works useful in illustrating the epistemological issues at stake in a philosophy of nature, though they do not speak directly to endangered species. See David Ehrenfeld, *The Arrogance of Humanism* (Oxford, 1978). Extremely useful as a study of how biology knows is Hans Jonas, *The Phenomenon of Life: Toward a Philosophical Biology* (Chicago, 1966). For a critique of our relationship with nature derived from the philosophy of Martin Heidegger, see Neil Evernden, *The Natural Alien: Humankind and Environment* (Toronto, 1985).

Another work develops a radical feminist response to nature: Susan Griffin, *Woman and Nature: The Roaring Inside Her* (New York, 1978). See also C. Merchant, *The Death of Nature: Women, Ecology and the Scientific Revolution* (New York, 1980).

My interpretation of endangered species in this book has been influenced by Michel Foucault. He speaks specifically of the formation of the modern view of nature in *The Order of Things: An Archaeology of the Human Sciences* (New York, 1970; originally published in France as *Les Mots et Choses,* Edition Galliard, 1966).

Two works have been particularly helpful in understanding nature and animals as social constructs: Harriett Ritvo, *The Animal Estate: The English and Other Creatures in the Victorian Age* (Cambridge, Mass., 1987), and Keith Thomas, *Man and the Natural World: A History of the Modern Sensibility* (New York, 1983). Another work that explores nature as metaphor is Roderick Nash, *Wilderness and the American Mind,* 3d ed. (New Haven, Conn., 1982).

On attitudes toward animals, see the bibliography by Stephen Kellert and Joyce K. Berry, *A Bibliography of Human/Animal Relations* (Lanham, Md., 1985).

Several works were indispensable for interpretations of the symbolic value of animals. Most impressive is Francis Klingender, *Animals in Art and Thought to the End of the Middle Ages* (Cambridge, Mass., 1971). A number of older sources are excellent: Alfred Newton, *A Dictionary of Birds* (London, 1896); D'Arcy Wentworth Thompson, *A Glossary of Greek Birds* (Oxford, 1936); William Norton Howe, *Animal Life in Italian Painting* (London, 1912); Arthur Collins, *Symbolism of Animals and Birds* (London, 1913); Ernest Ingersoll, *Birds in Legend, Fable and Folklore* (New York, 1923); Percy Ansell Robin, *Animal Lore in English Literature* (London, 1932); H. W. Seager, *Natural History in Shakespeare's England* (London, 1892); James E. Harting, *The Birds of Shakespeare* (1871; reprint, Chicago, 1965); George Boas, *The Happy Beast* (1933; reprint, New York, 1966).

I recommend two good articles on the question of the symbolic meanings of beasts, both in *American Imago*: Jacques Schnier, "The Symbolic Bird in Medieval and Renaissance Art," 9 (1952): 89–117, and Arthur Wormhoudt, "The Unconscious Bird Symbol in American Literature," 7 (1950): 173–181. See also Alexander Haggarty Krappe, "Guiding Animals," *Journal of American Folklore* 55 (1942): 228–246.

For short essays organized around individual species, excellent for reference, see two volumes by Beryl Rowland, *Animals with Human Faces: A Guide to Animal Symbolism* (1973) and *Birds with Human Souls: A Guide to Bird Symbolism* (1978). Though it is more limited in the range of species covered, see Angus K. Gillespie and Jay Mechling, eds., *American Wildlife in Symbol and Story* (1987). All three are published in Knoxville, Tenn.

The best scholarly summary of medieval views of animals as found in bestiaries is in Florence McCullogh, *Mediaeval Latin and French Bestiaries* (Chapel Hill, N.C., 1960). See also Helen Woodruff, "The Physiologus of Bern," *Art Bulletin* 12 (1930): 226–253, and Edward B. Ham, ed., "The Cambrai Bestiary," *Modern Philology* 36 (1939): 225–237.

Modern versions of older natural histories are available and make fascinating reading. See T. H. White, *The Book of Beasts, Being a Translation from a Latin Bestiary of the Twelfth Century* (New York, 1954; reprinted by Dover, 1984). Also interesting is *Topsell's Histories of Beasts*, edited by Malcolm South (Chicago, 1981), and *The Fowles of Heaven or History of Birds*, edited by Thomas P. Harrison and F. David Hoeniger, (Austin, Tex., 1972). Both were by the Elizabethan Edward Topsell.

Chapter 1: Hunger Makes the Wolf

The amount of material available on the wolf is vast, testimony to an abiding fascination with the wolf. Again, the following bibliography does not try to be exhaustive.

Several excellent books are available on the natural history of the wolf. The most readable, covering the wolf from a number of different perspectives, including natural history, is Barry Holstun Lopez, *Of Wolves and Men* (New York, 1978). The most thorough and definitive natural history of the wolf is L. David Mech, *The Wolf: The Ecology and Behavior of an Endangered Species* (Garden City, N.Y., 1970). Mech is perhaps the most commanding figure in the field of wolf biology. His most recent book, on his experiences with the wolves of Ellesmere Island, is *The Arctic Wolf: Living with the Pack* (Toronto, 1988).

For a readable account of wolves within a particular ecosystem, in this case Isle Royale, replete with observations and anecdotes, see Durward L. Allen, *Wolves of Minong: Their Vital Role in a Wild Community* (Boston, 1979); Allen draws upon a long and distinguished career studying wolves. Fred H. Harrington and Paul C. Paquet, eds., present a variety of interesting biological approaches to the wolf in *Wolves of the World: Perspectives of Behavior, Ecology, and Conservation* (Park Ridge, N.J., 1982). Also important is Michael Fox, ed., *The Wild Canids: Their Systematics, Behavioral Ecology, and Evolution* (New York, 1975).

Though dated, the classic early study of wolves is still very engaging: Adolf Murie, *The Wolves of Mount McKinley* (Washington, D.C., 1941; reprint, 1971). Murie was an early bedroll biologist, and his work still underlies more recent biological studies of the wolf. Also important in the transformation of the image of the wolf from a vicious beast to a member of an ecosystem is Aldo Leopold's essay "Thinking Like a Mountain," in *A Sand County Almanac, and Sketches Here and There* (Oxford, 1949).

Farley Mowat's fictional narrative with wolves, *Never Cry Wolf* (Boston, 1963), has become a de facto introduction to the wolf as predator for huge numbers of people, since the account was made into a movie. Most biologists bemoan the liberties he takes with the facts. Another introduction is Lois Crisler, *Arctic Wild* (New York, 1958).

Several popular articles also serve as entré to the biology of the wolf. Succinct information is given in Rolf O. Peterson, "The Gray Wolf," in Roger Di Silvestro, ed., *Audubon Wildlife Report 1986* (New York, 1986). Authoritative articles by L. David Mech are available in *National Geo-*

graphic. He wrote, with Durward Allen, "Wolf versus Moose on Isle Royale," 123 (February 1963): 200–219. Mech was the sole author of "Where Can the Wolf Survive?" 152 (October 1977): 518–537, and "At Home with the Arctic Wolf" 171 (May 1987): 562–593.

A large body of literature has emerged on the question of wolf control. I referred also to material on predator and prey relations, as well as studies on the behavior of wolves. The results of the study by the Alaska Department of Fish and Game can be found in a monograph by William C. Gasaway, Robert O. Stephenson, James L. Davis, Peter E. K. Shepherd, and Oliver E. Burris, *Interrelationships of Wolves, Prey, and Man in Interior Alaska* (The Wilderness Society *Wildlife Monograph* 84, July 1983). A popular version of the monograph can be found in William C. Gasaway, Robert O. Stephenson, and James L. Davis, *Wolf-Prey Relationships in Interior Alaska* (Juneau, Alaska Department of Fish and Game, Wildlife Technical Bulletin no. 6, 1983).

L. David Mech's studies on wolf-prey relationships can be found in several works. He is the coauthor, with Patrick D. Karns, of *The Role of the Wolf in a Deer Decline in the Superior National Forest* (St. Paul, Minn., North Central Forest Experiment Station, U.S. Department of Agriculture, Forest Service Research Paper NC-148, 1977). See also Mech's chapter "Predators and Predation," in L. K. Hall, ed., *White-tailed Deer Ecology and Management* (Harrisburg, Penn., 1984); his article "Productivity, Mortality, and Population Trends of Wolves in Northeastern Minnesota," *Journal of Mammalogy* 58 (1977): 559–574; and his more popular discussion of predator-prey relations and wolf control, "How Delicate Is the Balance of Nature?" *National Wildlife* 23 (February–March 1985): 54–58.

For a detailed recounting of the problems faced by the biologists in Alaska who were trying to understand the crash in the early 1970s, told in narrative style by a biologist involved, see Warren B. Ballard, "The Case of the Disappearing Moose," in three parts, *Alaska Magazine* (January, February, March, 1983). Results of another pertinent study are in Warren B. Ballard, Ted H. Spraker, and Kenton P. Taylor, "Causes of Neonatal Moose Calf Mortality in South Central Alaska," *Journal of Wildlife Management* 45 (1981): 335–342.

For wolves on Isle Royale, and their relationship to prey, consult the annual reports of the national park. I used Rolf O. Peterson, "Ecological Studies of the Wolves on Isle Royale, Annual Report—1983–84."

The work of Ludwig Carbyn offers a Canadian perspective on the management of wolves. His editorial "A Delicate Balancing Act," *Outdoor*

Canada 13 (June–July 1985): 9–11, suggests that, from the point of view of a biologist working in a national park, the controversy over wolves can be seen as hunters versus wolves. For his scientific studies, see "Management of Non-Endangered Wolf Populations in Canada," *Acta Zoologica Fennica* 174 (1981): 239–243, and "Wolf Predation on Elk in Riding Mountain National Park, Manitoba," *Journal of Wildlife Management* 47 (1983): 963–976.

The chapter by Guy E. Connolly, "Predators and Predator Control," in John L. Schmidt and Douglas L. Gilbert, eds., *Big Game of North America: Ecology and Management*, (Harrisburg, Penn., 1978), provides a helpful overview of this controversial biological issue.

Several articles treat the issue from the perspective of studies on the prey. For a focus on caribou, see several works by James L. Davis, with coauthors: "Demography and Limiting Factors of Alaska's Delta Caribou Herd, 1954–1981," *Acta Zoologica Fennica* 175 (1983): 135–137, and articles in E. Reimers, E. Gaare, and S. Skjenneberg, eds., *Proceedings of the 2d International Reindeer/Caribou Symposium, Røros, Norway, 1979* (Trondheim, Direktoratet for vilt og ferskvannsfisk).

Among the many excellent works on wolf behavior, I recommend several articles by Robert Stephenson. Two appeared in Fred H. Harrington and Paul C. Paquet, eds., *Wolves of the World*: "Wolf Movements and Food Habits in Northwest Alaska" and "Nunamiut Eskimos, Wildlife, Biologists and Wolves." Along with Warren B. Ballard and Rick Farnell, Stephenson coauthored "Long Distance Movements by Gray Wolves, *Canis lupus*," *Canadian Field-Naturalist* 97 (1983): 333.

See also Warren B. Ballard and James R. Dau, "Characteristics of Gray Wolf, *Canis lupus*, Den and Rendezvous Sites in Southcentral Alaska," *Canadian Field-Naturalist* 97 (1983): 299–302.

Considerable interest in the demography of American attitudes toward wildlife has developed recently, and a statistical sociology of contemporary attitudes toward wolves can be seen in Stephen Kellert, "Minnesotans and Wolves: A Survey," *Defenders* 60 (May–June 1985): 16–19.

The understanding of scientific ideas by their location within cultural history is currently one of the most vital fields in the sociology of knowledge. It has not spread, however, into our awareness of animals—that is, we still maintain the division between the scientific view of animals and the cultural views. Many sources might have been used in my discussion of the implications of objectivity and positivism in science. The references to Susanne K. Langer are from her classic study, *Philosophy in a New Key: A Study in the Symbolism of Reason, Rite, and Art*, 3d ed. (Cambridge, Mass., 1941; reprint, 1957). A more recent study shows how some of our more basic ideas about "nature," "science," and "life itself" are historically constituted, even though these ideas inform and influence what seem to be the most "objective" and "factual" inquiries: L. F. Jordanova, ed., *Languages of Nature: Critical Essays on Science and Literature* (New Brunswick, N.J., 1986).

For René Descartes, see *Discourse on Method and the Meditations*, translated and introduced by E. Sutcliffe (London, 1968).

Concerning the wolf in literature, myth, and psychology: For the Jungian interpretation of the wolf and the dark aspects of the sun, see Marie-Louise von Franz, *Alchemy: An Introduction to the Symbolism and Psychology* (Toronto, 1980). Shakespeare uses the wolf frequently as an image of wildness and appetite. The quotation I used from *Troilus and Cressida* is I.iii.121–124. Jacobean literature makes frequent use of the wolf as dangerous outcast. See especially John Webster, *The White Devil*, edited by J. R. Mulryne (Lincoln, Neb., Regents Renaissance Drama, 1969). For provocative use of the wolf as inner agent—the psychological werewolf—see Webster's *The Duchess of Malfi*, edited by Elizabeth M. Brennan (London, New Mermaid Series, 1987).

For a modern interpretation of lycanthropy, see Montague Summers, *The Werewolf* (Secaucus, N.J., 1966). Related is Richard Bernheimer, *Wild Men in the Middle Ages* (Cambridge, Mass., 1952).

For Charles Perrault's tale of Little Red Riding Hood, I used *Perrault's Complete Fairy Tales*, translated from the French by A. E. Johnson et al. (New York, 1960). Other representations of Little Red Riding Hood, as well as other fairy tales involving wolves, can be found in the Grimm brothers' *Grimm's Fairy Tales* (New York, 1944) and in Iona and Peter Opie, *The Classic Fairy Tales* (London, 1974). Also, for Jean de La Fontaine, Marianne Moore's translation, *The Fables of La Fontaine* (New York, 1954), is excellent.

Bruno Bettelheim offers a Freudian interpretation of Little Red Riding Hood in *The Uses of Enchantment: The Meaning and Importance of Fairy Tales* (New York, 1976). Beryl Rowland, in *Animals with Human Faces* (Knoxville, Tenn., 1973), provides literary, colloquial, and interpretive uses of the wolf.

For the references to medieval bestiaries, I drew from several sources. Florence McCullogh, in *Medieval Latin and French Bestiaries* (Chapel Hill, N.C., 1960), provides historical and textual commentary, along with a compilation of the many different images of the wolf in the bestiary tradition. For a complete example of the medieval bestiary, translated into

English, see T. H. White, *The Book of Beasts* (New York, 1954; reprinted by Dover, 1984). Edward Topsell's *Historie of Foure-Footed Beasts* (1607) can be approached in *Topsell's Histories of Beasts*, edited by Malcolm South (Chicago, 1981). Topsell also offers a unique interpretation of lycanthropy.

The quotation from colonial New England, "A waste and howling wilderness," is by Michael Wigglesworth (1662); it can be found in "God's Controversy with New England," Massachusetts Historical Society *Proceedings* 12 (1871–1873): 83–84.

Chapter 2: The Fall of a Sparrow

A standard life history of the dusky seaside sparrow is available in A. C. Bent's massive study *Life Histories of North American Birds* (New York, 1919–1968). See *Life Histories of North American Cardinals, Grosbeaks, Buntings, Towhees, Finches, Sparrows, and Allies; Order Passeriformes, Family Fringillidae*, 3 vols., edited by Oliver L. Austin, Jr. (Washington, D.C., 1968).

Several articles give an excellent overview of the loss of the dusky seaside sparrow, and they have a cumulative emotional impact. See Paul W. Sykes, Jr., "Decline and Disappearance of the Dusky Seaside Sparrow from Merritt Island, Florida," *American Birds* 34 (1980): 728–737. For Herbert W. Kale II's historical review of the loss of the bird, and the species' requirements for survival at the time, see "Endangered Species: Dusky Seaside Sparrow," *Florida Naturalist* 50 (February 1977): 16–21.

For a description of the surveys for the dusky, see Herb Kale II, "The 1980 Dusky Seaside Sparrow Survey," *Florida Field Naturalist* 9 (1981): 64–67.

Also fascinating to read are several other articles by Herb Kale II, as they trace the loss of the duskies from the point of view of a man actively engaged in saving them. See "A Status Report on the Dusky Seaside Sparrow," *Bird Conservation* 1 (1983): 128–132. Also, "The Dusky Seaside Sparrow: Have We Learned Anything?" *Florida Naturalist* 60 (Fall 1987): 2–3. Other articles by Herb Kale II in *Florida Naturalist* document the birds' changing circumstances, particularly "Duskies Transferred to Discovery Island," 56 (December 1983): 3.

For other reports on the seaside sparrow, see articles in the *Endangered Species Technical Bulletin Reprint* 2 (October 1985), and *Endangered Species Technical Bulletin* 12 (May–June 1987).

The figure I used on the number of species and subspecies that have gone extinct in the United States comes from Paul A. Opler, "The Parade of Passing Species," *Science Teacher* 44 (December 1976): 30–34.

The literary quotations are *Hamlet*, V.ii.219–220; Catullus, *Catullus*, edited by Elmer Truesdell Merrill (Cambridge, Mass., 1893; reprint, 1951), poems 2 and 3; and John Keats, *Keats: Poems and Selected Letters*, edited by Carlos Baker (New York, 1962), the letter dated 22 November 1817, to Benjamin Bailey.

Chapter 3: Carrion for Condors

The best general study of the natural history of the California condor may still be Carl B. Koford, *The California Condor* (New York, 1953). One of the National Audubon Society's monographs on American wildlife, it is dated, but full of interesting historical and biological information. Sandford R. Wilbur brings the study of the condor well into the 1970s with *The California Condor, 1966–1976: A Look at Its Past and Future* (Washington, D.C., U.S. Department of the Interior, North American Fauna, no. 72, 1978).

A careful review of the final phase of the work on the California condor in the wild is presented by two of the principal players in the events: Noel F. R. Snyder and Helen Snyder, "Biology and Conservation of California Condors," in Richard F. Johnston, ed., *Current Ornithology*, vol. 6 (New York, 1989).

For a perspective by another participant on the recovery team, see J. C. Ogden, "The California Condor Recovery Program: An Overview," *Bird Conservation* 1 (1983): 87–102. See also his article "The California Condor" in Roger Di Silvestro, ed., *Audubon Wildlife Report 1985* (New York, 1985).

For a readable, anecdotal perspective on the condor by a rancher who was actively involved in conservation, see Ian McMillan, *Man and the California Condor* (New York, 1968).

Several articles in the *Endangered Species Technical Bulletin* not only offer up-to-date information on the condor during its final stage in the wild but give a sense, too, of the ongoing efforts of biologists to save the last birds: "California Condor Population Grows by One," 13 (5, 1988): 1; "Captive California Condor Population," 12 (1, 1987): 3; Oliver H. Pattee, "The Role of Lead in Condor Mortality," 12 (9, 1987): 6–7; Robin B. Goodloe, "Recent Advances in the California Condor Research and Recovery Team," 9 (12, 1984): 8–10; and, by the same author, "Special Report: The California Condor, Building a Captive Breeding Flock," 10 (2, 1985): 5–6.

A number of excellent articles show how biologists got a grip on the problems facing the condors in the intensive recovery program in the 1980s. See especially the works by Noel Snyder, with coauthors: "Replacement

Clutching and Annual Nesting of California Condors," *Condor* 87 (1985): 374–378, and "Photographic Censusing of the 1982–1983 California Condor Population," *Condor* 87 (1985): 1–13.

An excellent series of popular articles on a wide variety of aspects of the California condor can be found in *Outdoor California* 44 (September– October 1983). The issue is dedicated to the condor and includes an article on the role of the condor in California Indian myths. Another interesting article is Dean Amadon, "Bare-Headed Vulture of California," *Living Bird Quarterly* 3 (1, 1984): 4–7.

There is an enormous literature about what Shakespeare, in *Titus Andronicus*, calls the "gnawing vulture of the mind." The notion of the vulture's conceiving without a mate is developed (along with other animals) in a fascinating article by Conway Zircle, "Animals Impregnated by the Wind," *Isis* 25 (1936): 59–130. For Horapollo, see *The Hieroglyphics of Horapollo*, translated by George Boas (New York, 1950). The quote from St. Ambrose of Milan is taken from his *Hexaemeron*, V, as quoted in Zircle. See also Beryl Rowland, *Birds with Human Souls* (Knoxville, Tenn., 1978). I also found Francis Klingender, *Animals in Art and Thought* (Cambridge, Mass., 1971), especially good on the various medieval traditions of interpreting nature.

I used an older edition for Oliver Goldsmith, *A History of the Earth and Animated Nature*, 2 vols. (London, 1853). Miguel de Unamuno's poem, "Este buitre voraz," comes from Carmelo Virgillo, L. Teresa Valdivieso, and Edward H. Freidman, eds., *Aproximaciones al estudio de la literatura hispánica* (New York, 1983). For the translation, I am indebted to Kay Garcia.

For Michel Foucault's analysis of power, see *The History of Sexuality, Vol. 1: An Introduction*, translated by Robert Hurley (New York, 1978). It was originally published as *La volenté de savoir* (Paris, 1976).

Chapter 4: A Panther in a Swamp

Compared to the literature on other high-profile endangered American species, relatively little has been published on the Florida panther.

For a current overview of the biology and conservation of the species, see Robert C. Belden, "The Florida Panther," in William J. Chandler, ed., *Audubon Wildlife Report 1988/1989* (San Diego, Calif., 1988). Jayde C. Roofe and David S. Maehr have published the results of some of their recent research in "Sign Surveys for Florida Panthers on Peripheral Areas of Their Known Range, *Florida Field Naturalist* 16 (1988): 81–104.

Other recent studies are Robert C. Belden, "Florida Panther Recovery Plan Implementation—a 1983 Progress Report," in S. D. Miller and D. D. Everett, eds., *Cats of the World: Biology, Conservation and Management* (Kingsville, Tex., Proceedings of the Second International Symposium, Caesare Klebery Wildlife Research Institute, 1986), and Steven A. Osofsky, "Panther Diary," *Natural History* 97 (April 1988): 50–54. A succinct summary of current work is in David J. Wesley, "The Florida Panther Recovery Program," *Endangered Species Technical Bulletin* 12 (1, 1987): 6–7.

Older studies of the Florida panther include C. B. Cory, *Hunting and Fishing in Florida* (Boston, 1896); O. Bangs, "The Florida Puma," *Proceedings of the Biological Society of Washington* 13 (1899): 15–17; R. Allen, "Notes on the Florida Panther, *Felis concolor coryi* Bangs," *Journal of Mammalogy* 31 (1950): 279–280; and Robert C. Belden, "Florida Panther Investigation— A Progress Report," in *Proceedings of the Rare and Endangered Wildlife Symposium* (Athens, Georgia Department of Natural Resources Technical Bulletin WL4, 1978).

For general articles on mountain lions, see Kenneth R. Dixon, "Mountain Lion," in J. A. Chapman and G. A. Feldhammer, eds., *Wild Mammals of North America* (Baltimore, 1982); K. R. Russell, "Mountain Lion," in J. L. Schmidt and D. L. Gilbert, eds., *Big Game of North America: Ecology and Management*, (Harrisburg, Penn., 1978); and S. P. Young and E. A. Goldman, *The Puma, Mysterious American Cat* (Washington, D.C., 1946).

An emotional book on the Florida panther, stressing psychological renewal as well as species renewal, is by James McMullen: *Cry of the Panther* (Sarasota, Fl., 1984).

Though I did not mention it in the chapter, I recommend Jean Conger, *The Velvet Paw: A History of Cats in Life, Mythology and Art* (New York, 1963).

Chapter 5: Sirens

Because of their gentle disposition, defenseless habits, and connections with mermaids, manatees have inspired a long bibliography of popular articles, of which a few of the best are listed below.

The best sources for scientific information on the manatee are Daniel S. Hartman, *Ecology and Behavior of the Manatee* (Trichechus manatus) *in Florida* (Pittsburg, Pa., American Society of Mammalogists, Special Publication no. 5, 1979), and Patrick M. Rose, "The West Indian Manatee," in Roger Di Silvestro, ed., *Audubon Wildlife Report 1985* (New York, 1985). Also excellent for scientific articles on the biology of the manatee is R. L.

Brownell and K. Ralls, eds., *The West Indian Manatee in Florida* (Florida Department of Natural Resources, proceedings of a workshop held in Orlando, Fl., March 27 to 29, 1978).

For a study of one of the most crucial issues facing the conservation of the manatee, see T. J. O'Shea, et al., "An Analysis of Manatee Mortality Patterns in Florida 1976–81," *Journal of Wildlife Management* 49 (1985): 1–11.

John E. Reynolds, in "The Semi-social Manatee," *Natural History* 88 (February 1979): 44–53, provides an excellent and accessible description of the behavior of the manatee. For a more technical discussion, see John E. Reynolds, "Aspects of the Social Behavior and Herd Structure of a Semi-isolated Colony of West Indian Manatees, *Trichechus manatus*," *Mammalia* 45 (1981): 431–451. See also the article by D. P. Domning, "Evolution of Manatees: A Speculative History," *Journal of Paleontology* 56 (1982): 599–619, for the history and origins of the creatures.

For a brief review of the early New World literature mentioning the manatee, see J. Baughman, "Some Early Notices on American Manatees and the Mode of their Capture," *Journal of Mammalogy* 27 (1946): 234–239.

Though dated and sometimes highly technical, an extensive bibliography of Sirenia is available: William K. Whitefield, Jr., and Sandra L. Farrington, *An Annotated Bibliography of Sirenia* (St. Petersburg, Florida Department of Natural Resources, Marine Research Laboratory, Florida Marine Research Publications no. 7, 1975).

The plight of the manatee and its conflict with humans have been described in a wide range of popular sources, and I mention only a few of the more prominent articles here. A succinct and factual account of the problems facing manatees is given in Thomas Baugh, "Man and Manatee: Planning for the Future," *Endangered Species Technical Bulletin* 12 (9, 1987): 7. For carefully documented research and exposition, see Faith McNulty, "Manatees," *The New Yorker* 55 (February 26, 1979): 83–89. See also several articles in *Oceans*, including John E. Reynolds, "Precarious Survival of the Florida Manatee," 10 (September–October 1977): 50–53 and Sue Douglas, "To Save a Vanishing Floridian," 15 (November–December 1982): 8–15.

National Geographic has published several articles on manatees: Daniel S. Hartman, "Florida's Manatees, Mermaids in Peril," 136 (September 1969): 342–353; Alice J. Hall, "Man and Manatees: Can We Live Together?" 166 (September 1984): 400–413; and Jesse R. White. "Man Can Save the Manatee," 166 (September 1984): 414–418.

Another good article is by Barbara Sleeper and Robert Rattner, "A Far Cry from a Sea Nymph," *Audubon* 88 (March 1986): 86–100.

Concerning the extinction of the Steller's sea cow, as well as the explorations of Steller, see Leonhard Stejneger, *Georg Wilhelm Steller, The Pioneer of Alaskan Natural History* (Cambridge, Mass., 1936). Stejneger also wrote the primary study of the species' extermination, "On the Extermination of the Great Northern Seacow (*Rhytina*)," *Bulletin of the American Geographical Society* 4 (1886): 317–328. See also C. Ford, *Where the Sea Breaks the Ice: The Epic Story of a Pioneer Naturalist and the Discovery of Alaska* (Boston, 1966). The best recent articles on the lost siren are two by Delphine Haley: "Saga of the Steller's Sea Cow," *Natural History* 87 (November 1978): 9–17, and "The Great Northern Sea Cow," *Oceans* 13 (September–October 1980): 7–11; and one by Robert McNully, "The Short Unhappy Saga of the Steller's Sea Cow," *Sea Frontiers* 30 (May–June 1984): 168–172.

A brief but interesting historical account of the manatee's association with the mythology of the mermaid is available in F. Smith, "Science and the Mermaid," *Sea Frontiers* 5 (May 1959): 74–82.

For an identification of the mermaid with Aphrodite and sea goddesses, see Robert Graves, *The White Goddess* (1948; emended and enlarged, London, 1962).

The book by Hans Jonas to which I referred is *The Phenomenon of Life* (Chicago, 1966). For Columbus's journal of his first voyage, I used *The Journal of Christopher Columbus*, translated by Cecil Jane (London, 1968). The mermaid appears in many more places than I discussed in the text, from Dante to Hans Christian Andersen. My quotations from *A Midsummer Night's Dream* are from V.i.2–29 and II.i.148–155. For T.S. Eliot, "The Love Song of J. Alfred Prufrock," I used *The Complete Poems and Plays* (New York, n.d.).

My references to the change in natural history at the end of the sixteenth century come from Keith Thomas, *Man and the Natural World* (New York, 1982), which offers a wonderfully detailed study of premodern ways of talking and thinking about animals, and from Michel Foucault, *The Order of Things* (New York, 1973), which is theoretical and, in its way, poetic.

Chapter 6: Guns and Parrots

Noel Snyder, James Wiley, and Cameron Kepler have prepared the most comprehensive monograph on the Puerto Rican parrot, *The Parrots of Luquillo: The Natural History and Conservation of the Endangered Puerto Rican*

Parrot (Los Angeles, Western Foundation of Vertebrate Zoology, 1987). These three biologists are the preeminent students of the Puerto Rican parrot in recent times, and their work is a major study of all biological aspects of the Puerto Rican parrot.

James Wiley has written two articles which provide a clear overview of the situation of the parrot and strategies for its conservation: "The Puerto Rican Parrot: (*Amazona vittata*): Its Decline and the Program for Its Conservation," in R. E. Pasquier, ed.,*Conservation of New World Parrots*, (Washington, D.C., Smithsonian Institution Press, International Council for Bird Preservation Technical Publication no. 1, 1980), and "The Captive Programme for the Endangered Puerto Rican Parrot, *Amazona vittata*," *Avicultural Magazine* 91 (1985): 110–116. His chapter "The Puerto Rican Parrot and Competition for its Nest Sites," in P. J. Moors, ed., *Conservation Management of Islands*, (ICBP Technical Publication no. 3, 1985) also provides an overview of conservation efforts for the parrot.

Excellent for understanding the work on the recovery of the Puerto Rican parrot are two articles by Noel Snyder in Stanley A. Temple, ed., *Endangered Birds: Management Techniques for Preserving Threatened Species*, (Madison, Wis., 1978). They are "Puerto Rican Parrots and Nest-site Scarcity" and, with J. D. Taapken, "Puerto Rican Parrots and Nest Predation by Pearly-eyed Thrashers."

A succinct work on the conservation of the Puerto Rican parrot is available in Kirk Horn, "The Puerto Rican Parrot," in Roger Di Silvestro, ed., *Audubon Wildlife Report 1985* (New York, 1985).

Several sources provide fascinating reading on the historical status and decline of the Puerto Rican parrot. Especially informative are two studies in the early twentieth century by Alexander Wetmore: *Birds of Porto Rico* (U.S. Department of Agriculture Bulletin no. 326, 1916) and *Scientific Survey of Porto Rico and the Virgin Islands* (New York Academy of Sciences, vol. 9, parts 3 and 4, 1927).

For the decline of the parrot in the twentieth century, see José A. Rodriguez-Vidal, *Puerto Rican Parrot* (Amazona vittata vittata) *Study* (Puerto Rico, Department of Agriculture and Commerce, Monograph no. 1, 1959). The unpublished Ph.D. dissertation by F. W. Wadsworth, "The Development of Forest Land Resources of the Luquillo Mountains, Puerto Rico" (Ann Arbor, University of Michigan, 1949), contains information not recorded in other historical sources. A researcher from the last century is also worth reading: E. C. Taylor, "Five Months in the West Indies, Part 2," *Ibis* 6 (1864): 157–173.

For Columbus's description of his early encounters with the Caribbean islands, natives as well as parrots, I used *The Voyages of Christopher Columbus, Being the Journals of His First and Third, and the Letters Concerning His First and Last Voyages*, edited by Cecil Jane (London, 1930; reprint, New York, 1970). The quotation of John James Audubon comes from his description of Carolina parakeets in *The Bird Biographies of John James Audubon*, edited by Alice Ford (New York, 1957).

For the loss of tropical forests, and the threat it poses to life on earth, see especially Norman Myers, *The Sinking Ark* (Oxford, 1979) and *The Primary Source* (New York, 1984). For personal reflections by a preeminent modern biologist, see Edward O. Wilson, *Biophilia* (Cambridge, Mass., 1984). Two excellent shorter treatments of the loss of the tropical forests, and its significance for North Americans, can be found in Ghillean T. Prance, "The Amazon: Paradise Lost?" in Les Kaufman and Kenneth Mallory, eds., *The Last Extinction* (Cambridge, Mass., 1987), and in Jack Connors, "Empty Skies," *Harrowsmith Magazine* 3 (July–August 1988): 35–45, which discusses the loss of North American songbirds because of the loss of the forests on their wintering grounds.

The ground-breaking work on the significance of islands in the biology of extinction is R. H. MacArthur and E. O. Wilson, *The Theory of Island Biogeography* (Princeton, NJ., 1967). I found useful several works on the loss of wildlife in Hawaii: J. Michael Scott and John L. Sincock, "Hawaiian Birds," in Roger Di Silvestro, ed., *Audubon Wildlife Report 1985* (New York, 1985); Cameron B. Kepler and J. Michael Scott, "Conservation of Island Ecosystems," in P. J. Moors, ed., *Conservation of Island Birds: Case Studies for the Management of Threatened Species* (Cambridge, England, ICBP Technical Bulletin Publication no. 3, 1985); and J. Michael Scott et al., *Forest Bird Communities of the Hawaiian Islands: Their Dynamics, Ecology and Conservation* (Lawrence, Kan., 1986).

An excellent popular article can be found in W. C. Gagne, "Hawaii's Tragic Dismemberment," *Defenders* 50 (June 1975): 461–470.

A huge bibliography exists on the philosophical issues I discussed, and it comprises both difficult and abstract reading. For the French anthropologist Claude Levi-Strauss, I used his autobiography *Tristes Tropiques*, translated by John and Doreen Weightman (New York, 1977; originally published by Librarie Plon, 1955). The most lucid summary of modern linguistics and its cultural implications that I have found is by Terence Hawkes. In *Structuralism and Semiotics* (Berkeley, Calif., 1977), Hawkes offers a superb bibliography, both for beginners and advanced students, on structuralism and semiotics. The famous quotation from Edward Sapir is in "The Status of Linguistics as a Science," in *Selected Writings of Edward Sapir in Language, Culture and Personality*, edited by David G. Mandelbaum (1949; reprint, Berkeley, Calif., 1968). For the role of language in self-knowledge, I used Jacques Lacan, "The Empty Word and the Full Word," in *The Language of the Self*, translated with notes and commentary by Anthony Wilden (Baltimore, 1968; originally published in Paris, 1956, in vol. 1 of *La Psychanalyse*).

The quotation on the dream of the parrot in Truman Capote is from *In Cold Blood* (New York, 1965).

Chapter 7: Green Eyes at Night

The rediscovery of the black-footed ferret sparked a renaissance of studies and research reports on the animals in the colony, as well as national media attention. The ferrets in that colony went extinct, however, before a definitive study on the natural history of the species could be completed.

For general natural history, describing ecology and behavior, perhaps the best work was published after the South Dakota ferret population was studied in the 1960s. The unpublished master's thesis by Conrad N. Hillman, "Life History and Ecology of the Black-footed Ferret in the Wild" (Brookings, South Dakota State University, 1968), describes field observations of the ferrets over sixteen months. A version of the thesis has been published as "Field Observations of Black-footed Ferrets in South Dakota," *Transactions of the North American Wildlife Natural Resources Conference* 33 (1968): 433–443. For the most comprehensive account of the ferret, with details of its natural history, see Robert F. Henderson, Paul Springer, and Richard Adrian, *The Black-footed Ferret in South Dakota* (Pierre, South Dakota Department of Game, Fish and Parks, Technical Bulletin no. 4, 1969). For additional information that summarizes many of their field experiences, see Raymond L. Lindner and Conrad N. Hillman, "The Black-footed Ferret," in a volume which they edited, *Proceedings of the Black-footed Ferret and Prairie Dog Workshop*, September 4 to 6, 1973 (Brookings, South Dakota State University, 1973).

Still interesting and good reading is Faith McNulty, *Must They Die? The Strange Case of the Prairie Dog and the Black-footed Ferret* (Garden City, N.Y., 1972). It studies the eradication of prairie dogs by government agents at the same time that other government agents were protecting endangered species like South Dakota's ferrets. The book was originally published as an article in *The New Yorker* 46 (June 13, 1970).

For an overview of the results from the Meeteetse colony of ferrets, perhaps the best source is the collection of articles in Stephen L. Wood, ed., *The Black-footed Ferret* (Brigham Young University, Great Basin Naturalist Memoirs no. 8, 1986).

In brief form, the article by Max Schroeder, "The Black-footed Ferret," in Roger Di Silvestro, ed., *Audubon Wildlife Report 1987* (New York, 1987), summarizes the status of the species in these post-Meeteetse days.

Though meant for young readers, an excellent general and readable account of the animal from firsthand experience, with abundant photographs, can be found in Denise Casey, *Black-footed Ferret* (New York, 1985).

Several accounts of the rediscovery of the ferrets in Meeteetse have been published in popular magazines. The account by one of the biologists active in the Meeteetse research and politics is Tim W. Clark, "Last of the Black-footed Ferrets?" *National Geographic* 163 (June 1983): 828–838.

For an interpretation that blames the collapse of the Meeteetse ferret colony on mismanagement, see David Weinberg, "Decline and Fall of the Black-footed Ferret," *Natural History* 66 (February 1986): 63–68. Another article in the same issue, by Louise Richardson, describes the winter search for signs of ferrets around Meeteetse.

For an interpretation that stresses not so much mismanagement as too much management, see Ted Williams, "The Final Ferret Fiasco," *Audubon* 88 (May 1986): 111–119.

The principal players in the catastrophe have begun presenting written interpretations. Tim W. Clark, in his article "Implementing Endangered Species Recovery Policy: Learning as We Go?" *Endangered Species Update* 5 (10, 1988): 35–42, uses the ferret recovery program as a case study in the bureaucratic problems inherent in wildlife management. E. Tom Thorne and Elizabeth S. Williams offer a different interpretation in "Disease and Endangered Species: The Black-footed Ferret as a Recent Example," *Conservation Biology* 2 (1, 1988): 66–74. They argue that the ferret was managed as well as possible given what was known at the time, and they retrace the history of events from the perspective of the governmental managers making the decisions. A popular version of this perspective is Tom Thorne, "A Future for Ferrets," *Wyoming Wildlife* 52 (October 1988): 20–27.

Tom Thorne, with Dave Belitsky and other coauthors, describes the disease of the ferrets in "Canine Distemper in Black-footed Ferrets (*Mustela nigripes*) from Wyoming," *Journal of Wildlife Diseases* 24 (1988): 385–398. See also Bob Oakleaf, "Why the Meeteetse Ferrets Had to Come In," *Wyoming Wildlife* 52 (March 1988): 28–33.

Tim W. Clark and coauthors have written several articles, highly statistical in method and indicative of one type of wildlife biology, that review various aspects of the Meeteetse ferret colony. See "Seasonality of Black-footed Ferret Diggings and Prairie Dog Burrow Plugging," *Journal of Wildlife Management* 48 (1984): 1441–1444, and "Black-footed Ferret Prey Requirements: An Energy Balance Estimate," *Journal of Wildlife Management* 47 (1983): 67–73.

Articles in the *Endangered Species Technical Bulletin Reprint* include two by Tim Clark: "Black-footed Ferret Recovery: Just a Matter of Time" 2 (June 1985) and "Black-footed Ferrets on the Edge," 3 (May 1987). An anonymous article goes back to an earlier time: "Only Known Ferret Population Receives Careful Attention" *Endangered Species Technical Bulletin* 8 (3, 1983): 5–8.

Audubon's account of his discovery of the black-footed ferret comes from his *Quadrupeds of North America*, written with the Rev. John Bachman, 2 vols. (New York, 1852–1854; originally published as *The Vivaporous Quadrupeds of North America*).

The literature on medicine pouches among the Plains Indians is extensive, if not very lucid. Much remains to be interpreted or understood about the way, for Indian cultures, animals confer power. Black-footed ferret pouches can be found in several Plains' museums: The Plains Indian Museum of the Buffalo Bill Museum in Cody, Wyoming, has one on display; Chief Plenty Coups Museum in Pryor, Montana, has four black-footed ferret pouches, which once belonged to Chief Plenty Coups; the Colter Bay Indian Museum in Grand Teton National Park has one skin sealed in plastic and stored.

Tim Clark details some of the uses of the pouches in "Some Relationships between Prairie Dogs, Black-footed Ferrets, Paleo-Indians, and Ethnographically Known Tribes," *Plains Anthropologist*, no volume number (1975): 71–74. For general reading on the search for power among Plains Indians, involving animals, see Robert H. Lowie, *Sun Dance of the Crow Indians*, edited by Clark Wissler (New York, Anthropological Papers of the American Museum of Natural History, vol. 16, part 1, 1921), and his *Indians of the Plains* (1954; reprint, Lincoln, 1982). Also useful, among many studies of the revival of Indian culture, was Fred W. Voget, *The Shoshoni-Crow Sun Dance* (Norman, Okla., 1984). I also found Alvin M. Josephy, Jr., *Now That the Buffalo's Gone: A Study in Today's American Indians* (Norman, Okla., 1982), eloquent background on what these people are trying to accomplish in their current history.

I know of only one poem about ferrets (at least, I think this poem describes a ferret): David Wagoner, "Burying a Weasel," which I found in *Defenders of Wildlife News* 40 (October 1965): 39. "Cat-whiskers, translucent claws . . . lithe deadliness," writes Wagoner of the dead ferret. But, he concludes, "My pasture was invented for dead animals."

Chapter 8: "So Ignoble a Leviathan"

Still a source of an incredible wealth of information, as well as being ex-hilirating in its poetry and its narrative, Herman Melville's *Moby Dick; or, The Whale* is like an encyclopedia that modern researchers continue to use. Though one must be careful in accepting what is offered as the biology of the whale, the novel is nevertheless amazing in its power and scope of vision, combining both information and imagination within its sweep. First published in 1851, *Moby Dick* is available in many editions, of course. I used Charles Feidelson, Jr., ed., (Indianapolis, Ind., 1964).

An excellent general source on all aspects of right-whale biology has been recently published with papers by leading researchers in the field: R. L. Brownell, Jr., P. B. Best, and J. H. Prescott, eds., *Right Whales: Past and Present Status* (Cambridge, England, International Whaling Commission Special Issue 10, 1986). See also Randall Reeves and R. L. Brownell, Jr., "Baleen Whales—*Eubalaena glacialis* and Allies," in J. A. Chapman and G. A. Feldhamer, eds., *Wild Mammals of North America: Biology, Management, and Economics*, (Baltimore, 1982). For estimates of the number of whales killed by the Yankee fishery, see Randall R. Reeves and Edward Mitchell, "Yankee Whaling for Right Whales in the North Atlantic Ocean," *The Whalewatcher* 17 (4, 1983): 3–5.

The article "The Northern Right Whale" in William Chandler, ed., *Audubon Wildlife Report 1988/1989* (San Diego, 1988), was written by re-searchers for the New England Aquarium: Scott D. Kraus, Marti J. Crone, and Amy Knowlton.

For papers on the pioneering studies on South Atlantic right whales, see several articles in Roger Payne, ed., *Communication and Behavior of Whales* (American Association for the Advancement of Science Symposia Series, 1983), especially on identification of and communication between right whales.

For a well-written account of the cetaceans of the world, with an evocation of the appeal of whales, see Stephen Leatherwood and Randall Reeves, *The Sierra Club Handbook of Whales and Dolphins* (San Francisco, 1983). It provides illustrations as well as accounts of the species' natural history, including right whales. I relied upon this work for some of the estimates of the populations of the different species of whales in the world, and those figures are, of course, subject to change over time. Also a good general reference, as well as a field guide, is Steve Katona, David Richardson, and Robin Hazard, *A Field Guide to the Whales and Seals of the Gulf of Maine* (Bar Harbor, Maine, 1975).

Several excellent books on whales in general are available. D. E. Gaskin's *The Ecology of Whales and Dolphins* (London, 1982) is controversial on the evolution of mammals, but excellent. The work of John Lilly did much to bring cetacaens to human attention. See, for example, *Man and Dolphin* (Garden City, N.Y., 1961) for his interest in the mind of the dolphin. A good introduction to cetacean biology is Theodore J. Walker, *Whale Primer* (San Diego, 1975). For general background on whales, see also Robert Burton, *The Life and Death of Whales* (1973; revised, London, 1980).

One of the most readable of books on whales is Robert McNally, *So Remorseless a Havoc: Of Dolphins, Whales and Men* (Boston, 1981). It is both intelligent and emotional. It also has an excellent, annotated bibliography.

Farley Mowat's *Sea of Slaughter* (New York, 1984) has powerful descriptions of the destruction of the life in the seas. For more recent developments in the fight against whaling, see Sidney Holt, "Conservation Update," *The Whalewatcher* 20 (2, 1986): 19–21. Richard Ellis, in "A Sea Change for Leviathan," *Audubon* 87 (November–December 1985): 62–79, offers an optimistic interpretation of the future for whales and explains how our appreciation for whales has changed in the last 15 years.

For an interesting and up-to-date examination of breaching in whales, with explanatory theories, see Hal Whitehead, "Why Whales Leap," *Scientific American* 252 (March 1985): 84–88.

For articles on right whales, see Randall Reeves, Scott Kraus, and Porter Turnbull, "Right Whale Refuge?" *Natural History* 83 (April 1983): 40–44, which recounts the discovery of the whales in the Bay of Fundy, and Charles Bergman, "Right Whales: Courting for Survival Along the East Coast," *Canadian Geographic* 108 (April–May 1988): 42–51.

For general background on whaling, I used Bill Spence, *Harpooned: The Story of Whaling* (New York, 1980); Irwin Shapiro and Edouard Stackpole, *The Story of Yankee Whaling* (New York, 1959); and Glover M. Allen, *Extinct and Vanishing Species of the Western Hemisphere, with the Marine Species of All Oceans* (Lancaster, Penn., Special Publication of the American Committee for International Wild Life Protection 11, 1942).

An excellent older source on right whales in the early period, with readable anecdotes, is Everett J. Edwards and Jeannette Edwards Rattray, *Whale Off: The Story of American Shore Whaling* (New York, 1932). It is the only extended account of this neglected aspect of American history. Also very interesting, with numerous quotations from early settlers, is George Francis Dow, *Whale Ships and Whaling, with an Account of the Whale Fishery in Colonial New England* (1925; reprinted by Dover, 1985). For an intriguing and detailed study, from shore whaling to the end of the golden age of New England whaling, see Edouard Stackpole, *The Sea-Hunters* (New York, 1953).

Several recent works have begun the process of uncovering the role of right whales in colonial American history. Most important is the research of Elizabeth Little. See especially "The Indian Contribution to Alongshore Whaling at Nantucket" (Nantucket Algonquin Studies no. 8, 1981) and, with Clinton Andrews, "Drift Whales at Nantucket: The Kindness of Moshup," *Man in the Northeast* 23 (1982): 17–38. Both include thorough bibliographies.

Also good is John A. Strong, "Shinnecock Whalers: A Case Study in Seventeenth-Century Assimilation Patterns," in William Cowan, ed., *Papers of the Seventeenth Algonquin Conference* (Ottawa, 1986).

A reconstruction of the Basque whaling off the Canadian coast can be found in James Tuck, "16th Century Basque Whaling Station," *Scientific American* 245 (November 1981): 180–184, and James Tuck and Robert Grenier, "Discovery in Labrador: A 16th Century Basque Whaling Port and Its Sunken Fleet," *National Geographic* 168 (July 1985): 40–71.

For the various legends of Moshup, see William S. Simmons, *Spirit of New England Tribes* (Hanover, N.H., 1986).

For an account of the myth of Moshup from a primary source, see Timothy Alden, Jr., "Memorabilia of Yarmouth," *Massachusetts Historical Society Collections* 1 (5, 1798): 54–60. For early descriptions of the whales, see Mourt's *Relation, Massachusetts Historical Society Collections* 1 (8, 1802): 23–39. The quotation by James Rosier is from "A True Relation of the Most Prosperous Voyage by Captain George Waymouth [1605]," *Massachusetts Historical Society Collections* 3 (8, 1843): 125–157. Accounts of shore whalers in action are from Hector St. John de Crèvecoeur, *Letters from an American Farmer* (New York, 1945; originally published in 1782).

Obed Macy is still interesting: *The History of Nantucket* (Boston, 1835). The quotation from Zaccheus Macy is taken from Elizabeth Little. The quotation of Cotton Mather is taken from Keith Thomas, *Man and the Natural World* (New York, 1983).

There is an exhaustive collection of art and literature on whales in a single beautiful source: Greg Gatenby, *Whales: A Celebration* (Boston and Toronto, 1983). The book is a marvel, testifying to the impact of the whale on the human imagination.

Rudyard Kipling's whale story, first published in 1902, is in *Just So Stories*, edited by Peter Levi (Hammondsworth, England, 1987). The other literary works to which I refer in the text are Carlo Collodi, *The Adventures of Pinocchio*, translated by M. A. Murray (London, 1911); Oppian's "Halieutica" in *Oppian Colluthus Tryphiodorus*, translated by A. W. Mair (Cambridge, Mass., Loeb Classical Library, 1928); D. H. Lawrence, *Studies in Classic American Literature* (New York, 1964; originally published in 1923) and *The Complete Poems* (New York, 1964); and Gary Snyder, "Mother Earth: Her Whales," in *Turtle Island* (New York, 1972).

For premodern naturalists on the whale, I found useful many of the sources mentioned in the "Introduction" section of this bibliography, especially H. W. Seager, *Natural History in Shakespeare's Time* (London, 1892). Also, an interesting scholarly article is by Cornelia C. Coulter, "The 'Great Fish' in Ancient and Medieval Story," *Transactions of the American Philological Society* 57 (1926): 32–50.

I gave only a brief overview of Michel Foucault's analyses of biology. His work is challenging, controversial, and stimulating in its critique of modern thought and culture. To read more of his study of biology, economics, and language, see *The Order of Things* (New York, 1973), especially Part II.

Chapter 9: Rumors of Existence

The best source of information on the life history and historical range of the bird is the monograph by James Tanner, *The Ivory-billed Woodpecker* (New York, National Audubon Society Research Report no. 1, 1942). The account is old now, and based on the study of only one population of the birds, but it remains the essential work on the species.

Another good source for natural history is Arthur Cleveland Bent, *Life Histories of North American Woodpeckers* (Smithsonian Institution U.S. National Museum *Bulletin* 174, 1939; reprinted by Dover, 1964). See also John K. Terres, *The Audubon Society Encyclopedia of North American Birds* (New York, 1980).

For the stories of the rediscovery of the ivory-billed woodpecker in the 1930s and famous controversial photographs of the woodpeckers in the

Atchafalaya, see George H. Lowery, Jr., *Louisiana Birds*, 3d ed. (Baton Rouge, 1974). The accounts convey his exhiliration in rediscovering ivory-billed woodpeckers in this century, as well as his frustration when, in the 1970s, he could get few people to believe his reports of this later and more controversial sighting.

For an important older account of the ivory-billed woodpecker by biologists who initiated Tanner's study, see A. A. Allen and P. P. Kellogg, "Recent Observations on the Ivory-billed Woodpecker," *Auk* 54 (1937): 164–184. Also interesting is E. A. McIlhenny, "The Passing of the Ivory-billed Woodpecker," *Auk* 58 (1941): 582–584.

My list of sightings of ivory-billed woodpeckers in this century comes largely from an unpublished report, "Ivory-billed Woodpecker Reports (U.S.)," compiled by Alisa Shull for the U.S. Fish and Wildlife Service, no date. The list is supplemented by personal communication with Jerome Jackson and Bruce Crider.

The political controversy over the funding, or lack thereof, by the U.S. Fish and Wildlife Service to support the preservation of and search for ivory-billed woodpeckers is covered in Michael Harwood, "You Can't Protect What Isn't There," *Audubon* 88 (November–December 1986): 108–123. He also examines the debate over whether the bird still exists in the United States. He believes it does.

Two interesting older articles on the search for the ivory-billed woodpecker are David Nevin, "The Irresistible, Elusive Allure of the Ivorybill," *Smithsonian* 4 (February 1974): 73–81, and Don Moser, "The Last: A Search for the Rarest Creature on Earth," *Life* 72 (April 7, 1972): 52–62.

John V. Dennis was one of the most controversial figures in the search for the ivory-billed woodpecker, embattled because of his firm conviction that he'd seen the species in Texas. For his articles, see "The Last Remnant of the Ivory-billed Woodpecker in Cuba," *Auk* 65 (1948): 497–507, and "The Ivory-bill Flies Still," *Audubon* 6 (November–December 1967): 38–45. A recent article of his compares the loss of the ivory-billed woodpecker with the resurgence of the pileated woodpecker: "Tale of Two Woodpeckers," *Living Bird Quarterly* 3 (Winter 1984): 18–21.

The material from the earlier American naturalists was derived from several sources. George Frederick Frick and Raymond Phineas Stearns give a careful analysis of Catesby's achievement, with quotations and illustrations, in *Mark Catesby: The Colonial Audubon* (Urbana, Ill., 1961). For Catesby's original work, see *The Natural History of Carolina, Florida, and the Bahama Islands*, 2 vols. (London, 1731–1743.)

My quotations from Alexander Wilson's account of the ivory-billed woodpecker come from *Wilson's American Ornithology*, edited by T. M. Brewer (Boston, 1840; vols. 1–7 originally published between 1808 and 1813; vols. 8 and 9, completed by George Ord, Wilson's friend and editor, in 1814). Another source is Alexander Wilson and Prince Charles Lucien Bonaparte, *American Ornithology; or, The Natural History of the Birds of the United States*, 3 vols. (London, 1832). For a definitive biography, see Robert Cantwell, *Alexander Wilson: Naturalist and Pioneer* (Philadelphia, 1961). More recent is Clark Hunter, ed., *The Life and Letters of Alexander Wilson* (Philadelphia, 1983).

For Thomas Nuttall, see *A Manual of Ornithology of the United States and Canada* (Cambridge, Mass., 1832).

For John James Audubon's description of the ivory-billed woodpecker, which I quoted in the text, see *Ornithological Biography*, 5 vols. (Edinburgh, 1831–1839). I used *The Birds of America* (New York, 1840–1844, "Miniature" edition) for a copy of his painting of the ivory-billed woodpecker. It was originally published in *The Birds of America* (London, 1827–1838, Double Elephant Folio). For his journals, see Maria Audubon, ed., *Audubon and His Journals*, 2 vols. (London, 1897; reprinted by Dover, 1960). See also the *Journal of John James Audubon Made during his Trip to New Orleans in 1820–1821* (Boston, 1929) and the *Letters of John James Audubon 1826–1840*, 2 vols. (Boston, 1930), both edited by Howard Corning. A detailed study of his life is found in F. H. Herrick, *Audubon the Naturalist: A History of His Life and Time*, 2 vols. (New York, 1917).

On the question of mass extinctions in modern times, see two papers in *The Dynamics of Extinction*, David K. Elliott, ed., (New York, 1985): Paul Ehrlich, "Extinction: What Is Happening Now and What Needs to Be Done," and Daniel Simberloff, "Are We on the Verge of a Mass Extinction in the Tropical Rain Forest?"

Also good are two papers in Les Kaufman and Kenneth Mallory, eds., *The Last Extinction* (Cambridge, Mass., 1986): David Jablonski, "Mass Extinctions: New Answers, New Questions," and David Ehrenfeld, "Life in the Next Millennium."

Additional figures on extinctions were derived from Paul A. Opler, "The Parade of Passing Species: A Survey of Extinctions in the United States," *Science Teacher* 44 (December 1976): 30–34.

For Indian use of ivory-billed woodpeckers, see A. M. Bailey, "Ivory-billed Woodpecker's Beak in an Indian Grave in Colorado," *Condor* 41 (1939): 164. Pipes with ivory-billed woodpecker parts (and some with feathers of the Carolina parakeet) can be found in the Milwaukee Public Museum. They are described in Alanson Skinner, "Ethnology of the Iowa Indians," Milwaukee Public Museum *Bulletin* 5 (44), June 12, 1926.

The folklore of the woodpecker and the herb of life is described in Pliny, *Natural History*, 10 vols., translated by H. Rackham (Cambridge, Mass., Loeb Classical Library, 1929). See vol. 3, book X.xx.

Concerning nearly vanished species, whose status as extinct or not is unclear, see the papers "The Bachman's Warbler" (by Paul B. Hamel) and "The Eskimo Curlew" (by J. Bernard Gollop), in William Chandler, ed., *Audubon Wildlife Report 1988/1989* (San Diego, 1988).

For articles on species I described in the section on common versus rare animals, see again the *Audubon Wildlife Reports*. Jerome Jackson's "The Red-Cockaded Woodpecker" is in *Audubon Wildlife Report 1987*, as is "The Black Bear," by Michael Pelton. See *Audubon Wildlife Report 1986* for the papers "The Spotted Owl" (by Eric Forsman and E. Charles Meslow) and "The Whooping Crane" (by James C. Lewis). *Audubon Wildlife Report 1985* has the paper "The Grizzly Bear" (by Chris Servheen). All three reports are edited by Roger Di Silvestro.

Chapter 10: The Back Door

For the trumpeter swan's brush with extinction and its unexpected "return," mentioned in the final chapter, see Charles Bergman, "The Triumphant Trumpeter Swan," *National Geographic* 168 (October 1985): 544–558.

INDEX

ABOUT THE AUTHOR

An active environmentalist, Charles Bergman has written extensively for *Audubon, Smithsonian, National Wildlife, National Geographic,* and *Orion Nature Quarterly*. He received his Ph.D. from the University of Minnesota and is a professor of English at Pacific Lutheran University. He lives in Tacoma, Washington.

Alaska Northwest Books™ proudly recommends several of its outstanding books on nature and the environment.

Living by Water: Essays on Life, Land, and Spirit, by Brenda Peterson
In the tradition of Henry Thoreau's reflections on Walden Pond to Annie Dillard's journal of life at Tinker Creek, novelist and environmentalist Brenda Peterson writes from her Puget Sound home about the ways in which water can shape a life and a philosophy.
144 pages, hardbound, $15.95/$19.95 Canadian   ISBN 0-88240-358-3

Two in the Far North, by Margaret E. Murie, illustrated by Olaus J. Murie
Margaret Murie describes her grand adventures on the Alaska frontier — in the gold-mining, dog-team era as a young bride, and then as trail mate and fellow wilderness explorer with her distinguished biologist husband, Olaus J. Murie. In this new edition, Murie extends her story to today's Alaska wilderness and parks issues, and Native land claims.
42 black-and-white drawings
396 pages, softbound, $12.95/$16.45 Canadian   ISBN 0-88240-111-4

Grizzly Cub: Five Years in the Life of a Bear, by Rick McIntyre
Grizzly Cub is the true story of a young bear's first five summers of life, as recorded in words and color photographs by Denali National Park ranger Rick McIntyre. The poignant episode that ensues when Little Stony becomes a "nuisance bear" dramatizes the conflict between the rights of park visitors to experience nature and the needs of wildlife to live unhampered by man.
56 color photographs
104 pages, softbound, $14.95/$18.95 Canadian   ISBN 0-88240-373-7

Guide to the Birds of Alaska, revised edition, by Robert H. Armstrong
"For those who plan to visit Alaska, this book is an obvious must, but even armchair birders will get satisfaction from reading [it]." —*The Canadian Field-Naturalist.*
With the addition of 32 species, this best-selling guide offers detailed information on all 437 species of birds found in Alaska.
Over 437 color photographs and watercolors
350 pages, softbound, $19.95/$24.95 Canadian   ISBN 0-88240-367-2

Alaska Northwest Books™

A division of GTE Discovery Publications, Inc.
P.O. Box 3007
Bothell, WA 98041-3007
1-800-343-4567